The Greening of the City

Urban parks are a much-loved feature of the city environment. However, our knowledge of the true scale of their impact remains uneven. Much work has been done on their origins and design features, but this book aims to extend this beyond the nineteenth century, examining the fuller flowering of these valuable spaces in the early decades of the twentieth century. Encompassing themes such as social and political usage, parks as employers and the dangers posed by such freely accessible spaces, the book examines a range of parks in cities such as Manchester, Salford, Liverpool, Leeds, Preston, Hull and Cardiff and challenges the prevailing myths about their meaning for their users. This study's timeframe spans almost 100 years of unprecedented social, cultural, political and economic changes and allows for the consideration of the expansion and commercialisation of leisure opportunities for the public. Urban parks played a significant role in this—the book places parks firmly in the context of the evolving city and examines the importance of green space to the urban citizen during this most fascinating of historical periods.

Carole A. O'Reilly is Senior Lecturer in Media and Cultural Studies at the University of Salford, Manchester, UK.

Routledge Studies in Cultural History

For more information about this series, please visit: www.routledge.com/
Routledge-Studies-in-Cultural-History/book-series/SE0367

The Greening of the City
Urban Parks and Public Leisure, 1840–1939

Carole A. O'Reilly

 Routledge
Taylor & Francis Group

NEW YORK AND LONDON

First published 2019
by Routledge
605 Third Avenue, New York, NY 10017

and by Routledge
2 Park Square, Milton Park, Abingdon, Oxon, OX14 4RN

First issued in paperback 2021

Routledge is an imprint of the Taylor & Francis Group, an informa business

Publisher's Note
The publisher has gone to great lengths to ensure the quality of this reprint but points out that some imperfections in the original copies may be apparent.

Library of Congress Cataloging-in-Publication Data
Names: O'Reilly, Carole A., author.
Title: The greening of the city : urban parks and public leisure, 1840–1939 / by Carole A. O'Reilly.
Description: New York : Routledge, 2019. | Series: Routledge studies in cultural history ; 73 | Includes bibliographical references and index.
Identifiers: LCCN 2019019995 (print) | LCCN 2019021614 (ebook) | ISBN 9781315866840 () | ISBN 9780415720663 (hbk) | ISBN 9781315866840 (ebk)
Subjects: LCSH: Urban parks—Great Britain—History—19th century. | Urban parks—Great Britain—History—20th century.
Classification: LCC SB484.G7 (ebook) | LCC SB484.G7 O74 2019 (print) | DDC 363.6/80941—dc23
LC record available at https://lccn.loc.gov/2019019995

ISBN 13: 978-1-03-209244-7 (pbk)
ISBN 13: 978-0-415-72066-3 (hbk)

Typeset in Sabon
by Apex CoVantage, LLC

To all who have been involved with or who have worked in public parks for the good of others

Contents

Figures

Abbreviations

HHC   Hull History Centre, Hull
LA   Liverpool Archives, Liverpool Central Library
LRO   Lancashire Record Office, Preston
MA   Manchester Archives, Manchester Central Library
SA   Salford Archives, Salford Local History Library
WYAS West Yorkshire Archives Service, Leeds

Acknowledgements

Any book such as this relies on the support and contributions of many people. This one is no different. My editor at Taylor and Francis, Max Novick, was a tower of positivity and encouragement, especially when the vicissitudes of life delayed the completion of the manuscript.

My PhD supervisor at Manchester Metropolitan University, Professor Alan Kidd, was a superb advisor, copyreader and general cheerleader for this project. Colleagues at the Directorate of Journalism at the School of Arts and Media, University of Salford, were a constant source of encouragement and inspiration.

There are too many libraries and archives to thank all individually, but I would like to mention Manchester Archives and Local Studies, Salford Archives, West Yorkshire Archives (Leeds), Liverpool Archives and Hull History Centre. Many individuals provided me with assistance—Rosie James, Principal Landscape Officer at Cardiff County Council, Gemma Nelson and Kathleen Hamilton of the Theatrical Management Association and Jenifer White at Historic England, all of whom contributed more than they know. The following were generous with their time and advice in sourcing the images: Tim Horning of University of Pennsylvania Archives and Records Center, Duncan McCormick of Salford Local History Library, Jarvis Gurr at Historic England, Julia Skinner at the Francis Frith Collection, Sally Hughes at Leeds Central Library, Louise Hunt at Glamorgan Archives and Colette Heavey at the Friends of Heaton Hall.

Tim Pettigrew invited me to his home and generously opened the Pettigrew family archive for me. Paul Rabbitts shared his expertise of park management and landscape design with me while Dr Katy Layton-Jones offered invaluable advice and support.

Parts of Chapter 2 are derived from articles published in *Urban History* (February 2013) © Cambridge University Press (available online: https://doi.org/10.1017/S0963926812000673) and *Landscape History* (October 2017) © the Society for Landscape Studies (available online: https://tandfonline.com/doi/abs/10.1080/01433768.2017.1394066).

Parts of Chapter 5 are derived from an article published in the *International Journal of Regional and Local History* (November 2013) © Taylor and Francis (available online: https://doi.org/10.1179/2051453013Z.0000000009).

1 Understanding Urban Parks

Introducing Urban Parks and Public Leisure

> Broad-minded, far-seeing public authorities appreciate the fact that the real assets derived from the provision of all pastimes in their parks are not monetary in character, but are the enhanced health and happiness of the community.
>
> (W. Pettigrew, 1937, p. 101)

Public parks are one of the most cherished elements of the British town and city of the twenty-first century. Their impact on public health, social cohesion, education and civic pride is often taken for granted. Yet, we know little about their origins, development and contribution to urban living as a whole. Parks have been host to sporting activity, political demonstrations, criminal behaviour and a wide diversity of civic and local initiatives since their inception in the 1840s. They have much to tell us about how life in the urban environment developed and thrived in spite of overcrowding, poverty and deprivation. They have impacted on our health, leisure opportunities and how we perceive our cities, their inhabitants and ourselves.

While public parks are most often associated with the Victorian period, their natural precursors were the public gardens of the seventeenth century (Taigel and Williamson, 1993). These spaces placed an emphasis on public walking and were based on the designs of private, aristocratic gardens. In many cases, no admission fees were charged, but income was often derived from the sale of refreshments (Taigel and Williamson, 1993). London's Royal Parks were opened to the public from the reign of Charles I. Provision was made in these parks for elite pursuits, such as racing and deer hunting (Taigel and Williamson, 1993). The upper classes also had access to private commercial pleasure grounds for entertainment. Public walkways had been provided in many British towns and cities since the seventeenth century, usually on the banks of a river or on the outskirts of the town. These were often simple tree-lined avenues where people could walk and be seen in public and were mainly frequented by polite society (Borsay, 2010).

The word 'park' was initially used to refer to the deer park beyond the formal gardens that abutted a country house. Parks were originally used for the practical purposes of deer hunting, grazing and providing food for the consumption of the residing family (Williamson, 1998). This definition of a park was later expanded in the eighteenth century to describe a landscape park, which referred to an open expanse of land with occasional clumps or belts of trees that was designed to provide a view for the owner or visitor. The aim of this kind of landscape was to demonstrate the wealth and power of the owner and to create a space that appeared 'naturally occurring' to the spectator. These landscapes were designed by architects like William Kent, Lancelot 'Capability' Brown and Humphrey Repton. As Williamson points out, these parks were the '*sine qua non* of true gentility', but they also represented a contrast to the urban landscape—secluded, private and rural (Williamson, 1998, p. 85). This was one aspect that was to remain important in defining the first public parks in Britain.

In many ways, the early municipal parks were inspired by the aristocratic connotations of the word 'park'. Many were designed by those who had worked on aristocratic estates, such as Joseph Paxton (Derby Arboretum, Derby). Some parks derived from these estates themselves as the landed class sought to sell them off when threatened by urban growth. Such estates were often acquired at a reasonable price by local authorities anxious to buy expanses of land for park-making. The 600 acres of Heaton Park cost Manchester City Council £230,000 in 1901 (O'Reilly, 2013).

The connection between the physical landscapes of the aristocracy and what were often referred to as 'peoples' parks' is a significant one. Many former estate buildings were converted into art galleries and museums and, as many already had elaborate planting schemes in place, they formed a simple route to an enjoyable public space for recreation. The kinds of pursuits undertaken in the early public parks also took their inspiration from the elite leisure interests of the wealthy—archery, cricket and tennis were among the most common sporting activities in the Victorian park and many of these sports remained popular until well into the twentieth century.

This chapter sets out to explore and discuss some of the diverse motivations behind the genesis of the parks movement in Britain. It examines the key concepts that underpinned that movement and gave it shape and direction—public health, moral authority, civic pride, social control and public leisure. It seeks to understand not just the general direction of this development but the differential trajectories followed in different English cities and towns at diverse times of their parks development.

The main themes of the book will now be outlined, followed by a note on the sources used for the study.

Parks and Public Health

The initial impetus for the establishment of public parks in Britain stemmed from a broad, cross-party consensus about public health. The 1833 Select Committee on Public Walks reported to Parliament that many British cities were in danger of condemning their poorer populations to shorter lives due to the lack of open space. Many writers have interpreted this as an attempt by the middle classes to extend further their social control of the working classes by developing spaces in which supervised physical exercise might be undertaken (Taylor, 1999 and Wyborn, 1995). This middle-class moral imperialism has been termed 'rational recreation' and it has been offered as an explanation for other improving initiatives of the Victorian period, such as libraries and museums.

Motivated by fears about working-class leisure habits like drinking and gambling, the earliest attempts at introducing open spaces into the Victorian city resemble a direct solution to this issue. However, both the motivations for the introduction of parks into the British city and the responses to the problematising of working-class culture conceal the more subtle and covert and often diverse reasons behind the emergence of the public park in the 1840s. Some cities, such as Leeds, came later to parks than others, and there were by no means similar motivations for their establishment in all British cities.

Offering more general explanations for the introduction of these precious spaces becomes problematic, as does the examination of the development of many constituent parts of urban life at this time. However, many cities and their municipal authorities were guided by broadly similar instincts when debating the establishment and funding of public parks and the controversies that often surrounded this issue were similar in nature and in tone. This may be attributed to the fact that, during much of the Victorian period, the social composition of such municipal authorities was homogenous—not just the 'hard-headed shopkeepers' of Beatrice Webb's rather harsh characterisation but also coalitions of pragmatic and benevolent civic-minded individuals, each motivated by a concern for their fellow citizens and a desire to bestow their legacy on the urban landscape (Kidd, 2006, p. 148). Public parks quickly became a difficult entity to oppose from a political perspective, and we therefore find local politicians of all backgrounds and parties often united in their support. This support was generally maintained after their establishment, when the realities of the costs of their upkeep became all too apparent. Much hope and investment in the future were often embodied in these spaces, not all of which were repaid, at least immediately.

It is not insignificant that it was a concern for public health that was the initial reason for the interest in adding public parks to the British civic landscape. The 1833 Select Committee on Public Walks heard evidence from representatives of many overcrowded and polluted cities, such as Manchester, who emphasised the poor quality of the urban environment and the consequent impact on the health of the population, especially the poor. The Industrial Revolution had made Britain one of the world's wealthiest and most powerful nations, but its legacy for the health of the public was more dubious. Social reformers, such as Edwin Chadwick and Henry Mayhew, later provided their own ample evidence of the living conditions endured by the urban poor, but much of the evidence presented to the 1833 committee was anecdotal in nature. Nonetheless, it was sufficiently persuasive for the committee to recommend that public walks be established to promote 'healthful exercise', as proposed by Robert Slaney (1791–1862), Liberal MP for Shrewsbury. He argued that the enclosure of land by private landlords had deprived many people of the opportunity to take exercise in the open air, away from the polluted air of their homes (Hansard, 1860, column 1288). Similarly, the 1840 Select Committee on the Health of Towns (chaired by Slaney) lamented the deplorable impact of poor housing on public health (Hunt, 2004).

The city was considered to be a 'body' in its own right. Just as a physical body could become infected by disease, so it was thought that the body of the city could be harmed by the presence in it of the poor and the marginalised (often ethnic minority immigrants, such as the Irish, were characterised in this way). This 'anatomical approach to the city' prevailed during the mid-nineteenth century and formalised a medical model to resolving social problems (Poovey, 1995, p. 74). Conceiving of the city as a social body resulted in approaching social problems as a kind of disease, with an emphasis on mutual interest (Poovey, 1995). This was reflected in the writing of nineteenth-century social reformers, such as James Phillips Kay, whose 1832 pamphlet, *The Moral and Physical Condition of the Working Classes in Manchester*, argued that the condition of the poor affected all social classes. Kay was a physician who believed in a link between moral and physical health. 'The state of the streets', he wrote, 'powerfully affects the health of their inhabitants' (1832, pp. 14–15).

This debate resulted in what is often referred to as the 'condition of England question', which focused the minds of writers such as Thomas Carlyle (Levin, 1998, p. 42). Carlyle also believed that cities damaged the organic bonds between individuals and represented a crisis of social relations that threatened all of society—'The condition of the great body of people in a country is the condition of the country itself' (Swift, 2001, p. 69). Thus, improving the physical conditions in which the majority of the working class lived would result in a similar advance in their moral behaviour. Such thinking laid the foundations for inquiries like the 1833 Select Committee on Public Walks and a determination to utilise access to open space for its curative and regenerative powers. The lack of availability of such spaces in crowded cities such as the ones that feature in this study continued to act as a deterrent to the consistent and regularised development of a sufficient number of urban parks.

One of the most enduring images of the public park is as a lung or a green lung for the overcrowded and polluted city. This image was especially prevalent during the Victorian period, which was also the zenith of the urban park. Most commonly associated with diseases such as consumption (tuberculosis) and pleurisy, many lung problems resulted from breathing dirty city air, living in overcrowded housing and working in heavy industries, such as coal mining and bleaching, which were all cornerstones of the nineteenth-century economy.

Lungs were especially vulnerable to smoky and polluted cities. The co-opting of the word 'lung' to describe a public park was intended to stress the rebalancing effect of an open space on the body of the citizen, enabling him or her to breathe more freely. If cityscape is the totality of the (civic) body, then the urban park represents a vital organ, which contributes to the healthy functioning of that body (and, by implication, the healthy mind). The interdependence of the municipal components of the city is made evident, and the case for their provision becomes common sense and therefore difficult to challenge.

The origin of the image of the park as a lung has been variously attributed to British prime minister William Pitt (Lord Chatham), Central Park designer Frederick Law Olmsted and Le Corbusier, the modernist architect and urbanist. In 1812, Mr Windham speaking in Parliament about Hyde Park attested that 'it was a saying of Lord Chatham that the parks were the lungs of London' (Cobbett, 1812, p. 1124). Urban trees and even certain roads and highways were also periodically referred to as lungs. Mosley suggests that it quickly became apparent that the trees that had been planted in cities such as Manchester could not withstand the extent of smoke pollution in that city (Mosley, 2001). Certain species of trees that could survive, such as poplar, ash and elder, were identified from the mid-nineteenth century (Mosley, 2001). These ecological challenges will be discussed more fully in Chapter 6. Public health reformers embraced the positive imagery of the park as lung to reinforce the need for public open space and to emphasise the rural aspect of the park in the city. This idealist tradition of anti-urbanism is significant when considering that many parks were located close to industrial areas and suffered from the same pollution as other parts of the city.

Nevertheless, the power of the idea of the park as lung should not be underestimated. It endured in much writing about urban parks throughout the twentieth century. It was challenged by Jane Jacobs only in the 1960s, who wrote that the park as lung was 'science fiction nonsense' (Jacobs, 1964, p. 101). She exposed the myth behind which the urban park and its assumed positive impact on the city environment lay and questioned the long-accepted causal link between the existence of city parks and their contribution to public health. In so doing, she drew our attention to the dangers of failing to develop a fully critical approach to these spaces and of treating all public parks as if they were the same. The

small patch of wasteland near the densely packed housing is not necessarily comparable to the large, carefully landscaped flagship urban park
in the suburbs. As Conway (1991) has remarked, there were continual
tensions between the park as a designed space and the provision of amenities for visitors.

Indeed, any evidence about the ability of urban parks to improve public health is elusive. In Liverpool, as in many British cities, epidemics
of cholera and typhus continued throughout the nineteenth century and
most of the urban poor did not have access to the city's three parks—
Stanley Park, Newsham Park and Sefton Park (Layton-Jones and Lee,
2008). Many of these parks suffered from the effects of pollution, the
Illustrated London News noting that the view from the terrace at Stanley Park was 'commonly obscured by the smoke of the factory districts'
(Layton-Jones and Lee, 2008, p. 35). Figure 1.1 gives an indication of the
extent of the polluted environment of nineteenth century Liverpool. As
late as 1897, direct links were still being made between parks and urban
death rates. Debating the purchase of the Buile Hill estate in Salford in
1897, Dr Quine drew a direct comparison between Salford's high death
rates and the city's 'lowest acreage of parks and open spaces' (*Manchester
Evening News*, 1897, p. 2).

An examination of the purported links between the provision of urban
parks and attempts to improve public health allows us to study the emergence of ideas about what is meant to be healthy during a period when

Figure 1.1 The smoky Liverpool skyline in 1890.

Source: © The Francis Frith Collection.

the body was often considered to be an instrument of sin. The gradual alteration in this view and its replacement by a sense of pride in a fit and healthy body had implications for the kinds of activities in which people indulged in many parks. The earliest often had simple gymnasia, featuring basic physical exercise equipment. This was later replaced by open-air physical exercise classes and by amenities such as running tracks where both men and women could improve their fitness and display their physical health. The growing acceptance of physical exertion as a key part of fitness for both men and women extended beyond a crude quest for national efficiency and reflected a new desire to display physical prowess in a public place. The new discourse of physical culture emphasised good health as a civic duty and a belief that the body could and should be improved and perfected (Zweiniger-Bargielowska, 2010).

A series of legislative initiatives during the early decades of the twentieth century focused greater attention on the acquisition of land for playing fields and recreation. These are studied in more detail in Chapter 5. The 1925 Public Health Act introduced the possibility of using parks to produce revenue from leisure and entertainment activities (Public Health Act, 1925, section 69). The Act enabled local authorities to rent out portions of public parks to local cricket and football clubs and to charge the public for admission to watch matches. The year 1925 also saw the creation of the National Playing Fields Association, especially focused on local communities' access to green open space. These ideas continued to be relevant until well into the 1930s. They were reinforced by the 1937 Physical Training and Recreation Act, which enabled the creation of physical training centres in cities. Such an emphasis on the improvement of physical fitness in the general population resulted in improved living standards and greater longevity, in which the existence and uses of urban parks played a part.

However, this was also a period when the use of parks for leisure and recreation was undergoing rapid change. Not only were parks under intense pressure to provide ever more diverse ranges of activities, but also they had to compete with newer and more attractive forms of commercial leisure, such as cinemas, theatres and dance halls. People were finding outlets for their leisure pursuits in very different ways. Central government also took steps to rein in the spending and power of local authorities by exerting greater control over their finances. This had the effect of limiting the ability of the municipalities to enact their own local leisure agendas (Stevenson, 1984).

From Social Control to Normative Behaviour in the Public Park

From their inception, the decisions made about public parks were political in nature. Though not uncontroversial, most local authorities accepted the arguments about the need for public space for recreation, although different towns and cities articulated this in varying ways and with different emphases. Whatever the circumstances, the political will and momentum had to be consistent and committed to sustain the development through the negotiations with landed families or other sellers all too willing to sell, but not at any price.

While it was difficult from a municipal point of view to be opposed to public parks, there is evidence of some concerted opposition from local ratepayers' groups in many British cities (Hunt, 2004). These groups were not opposed to the concept of public recreation per se, but to the contribution of parks to rising rate levels. In many cases, these groups were the only obvious opposition to the decisions made about parks by local authorities and they were frequently to be proved correct in their predictions about their eventual cost, both of purchase and maintenance.

In order to understand local municipal agendas and priorities, it is necessary to examine the kinds of arguments put forward by local authorities when discussing the need for public parks. A variety of different factors were usually considered—the location of the proposed park (generally confined to whether it was within the municipal boundary), the price of acquiring and developing the land, the nature of the land itself and its relationship to public transport infrastructure and developing suburban areas, the current state of public health in the city or town, the existence of the public parks in neighbouring cities and towns, civic competitiveness between neighbouring towns and cities, the perception of the nature of social problems in the area and the degree to which there was a consensus about whether public recreation was an appropriate response to these problems. The degree to which such a consensus existed across political party lines, the desire to address the problem of social order and the desire to provide spaces where different social classes could congregate together all conspired to drive the impetus to acquire land for public parks.

Many local authorities were hampered in their desire for enterprise by factors intrinsic and external to themselves. Redford (1939) has noted that Manchester City Council lacked a co-ordinated approach to developments within its boundaries and was able to act only rather opportunistically, while Ernest Simon remarked that Manchester remained polarised between those who wished to develop municipal services indefinitely and those who were more cautious about the burden on the ratepayers (Simon, 1926). Other city councils were limited by their proximity to such municipal power centres as Manchester. Garrard (1977) has observed that Salford's municipal leaders had a limited ability to act independently of Manchester, implying that they were hampered by their position as an ancillary or adjunct city.

Undoubtedly, there were also subtler forces at work. It was not uncommon, for instance, for parks committees to receive deputations and signed memorials in support of the purchase of land for recreation. These documents tended to be organised by influential ratepayers anxious to benefit from the increased land values that often resulted from the establishment of a park in a neighbourhood. Often designating themselves as a 'citizen's committee', these were citizens of a very particular kind—local businessmen, professionals, senior clergy and philanthropists. Such a committee submitted a memorial from 2,000 ratepayers to Manchester City Council in 1880 when the city was considering buying Manley Park. The mayor of Manchester, Alderman Patteson, urged the city council to receive a deputation from the citizen's committee, 'with a clear understanding that in doing that they did not commit themselves in any way whatever' (*Manchester Courier*, 1880). It is difficult to speculate with any certainty about the impact of such deputations and memorials, but the fact that they were a regular occurrence suggests that they were perceived as a successful method of exercising a form of subtle social pressure on the civic authorities.

The most surprising omission from the list of arguments in favour of urban parks is public demand (or lack of it). It has been suggested that there was little connection between public demand for recreational facilities and their introduction (Meller, 1976). However, the success of some municipal bodies in establishing a system of public subscription for their public parks suggests that such public demand did exist and was responsive enough to the idea of public parks to encourage the public to donate often substantial sums of money to their development. While the motivations of the individuals who did donate in this manner are hard to assess given the lack of evidence, we can speculate that they may have done so because they had the means available and were driven by a kind of local noblesse oblige, which impelled such acts. Public subscriptions funded libraries, art galleries, working men's reading rooms and other charitable endeavours and were an important and visible signal of social power and influence.

One of the most pressing concerns about the establishment of public parks was how they were to be regulated. Fear of large crowds of people in an easily accessible environment was driven by incidents such as the Peterloo massacre at a Chartist protest in St Peter's Field, Manchester in 1819. Eleven people were killed by the Manchester Yeomanry, dispersing the meeting (Kidd, 2006). Further Chartist demonstrations took place on Kersal Moor in Salford in 1839, Independent Labour Party meetings in Boggart Hole Clough Park in 1896, and suffrage gatherings in 1908 and demonstrations by the unemployed in front of the town hall in Albert Square in 1908 and again during the 1930s. Urban open space has, therefore, provided an opportunity for the citizens to express political views and to reclaim the landscape for themselves. These issues are outlined in more depth in Chapter 3. There is some evidence, however, that different groups used different kinds of public spaces for political purposes. It was more common for public parks to be used for organised demonstrations by particular groups, such as the Independent Labour Party, the Women's Social and Political Union and the United Kingdom Alliance, a temperance movement. Streets and squares were more commonly used for more spontaneous protests.

Such was the impact of the Chartist demonstrations that the 1844 Manchester Borough Police Act suggested that all public parks in the city should be situated at least five miles from the city centre to 'keep trouble out of town' (Manchester Borough Police Act, 1844). This was prompted not merely by the possibility of unregulated political meetings but also by fears that these new open spaces could be used for immoral purposes, thus rendering their original intentions to improve physical and moral health moot.

Indeed, it is possible to assert that Chartist activity provided an impetus for the development of public parks in many British towns and cities. While it remains challenging to evidence this directly, it is the case that there is a correlation between the main Chartist strongholds found in the manufacturing districts of Lancashire, Cheshire, the West Riding of Yorkshire and the east Midlands and the fact that by 1852 there were public parks in Derby (1840), Sheffield (1841), Bradford (1846), Manchester (1846), Salford (1846), Preston (1847), Darlington (1849) and Nottingham (1852). By the time of the 1842 Chartist Convention in London there were 43 Chartist localities in the metropolis, and it is possible that Bonner's Fields, in the East End of London, was chosen as the site for Victoria Park (1845) because it was a traditional place for the gathering of large crowds for meeting and rallies (Carter, 2011).

One of the earliest attempts to control access to public space was enclosure. Preston Corporation enclosed 100 of the 400 acres of Preston Moor in 1833, despite the protestations of the local Freemen who enjoyed rights of pasturage and other privileges there (Pollard, 1882). The enclosed area was developed to include an avenue and a carriage drive, and eventually sporting facilities, such as bowling and cricket, were added. In Cardiff, the Bute Trustees took steps to restrict public access to Cooper's Fields and the Castle Green, near Cardiff Castle in 1855, for reasons of privacy (A. Pettigrew, Volume 1, 1926, p. 31). The land had been widely used by the public for rambling along the well-established tracks. It took until 1858 for the Local Board of Health to set up a committee to investigate the loss of amenities for public walks in Cardiff, perhaps out of deference to the Bute family. The inquiries failed to establish the existence of any public rights of way with respect to the land and the matter was dropped (A. Pettigrew, Volume 1, 1926, p. 33).

The concept of social control has been deployed when studying the impact of public parks on the urban environment more generally. Most commonly, it has been utilised as a component of rational recreation and envisaged the park as an implement of middle-class moral imperialism. Put more positively, the parks were supposed to offer alternative 'rational' recreation to absorb the leisure time of the new working class of the industrial towns. Similar arguments have been made about the social role of museums, libraries and art galleries (Hill, 2001). The assumption underpinning these facilities as mechanisms of social control is to offer alternative rational recreation to absorb the leisure time of the new working class of the industrial towns and cities. The impact of Chartism is often used to embody the alleged middle-class fear of the urban mob and the subsequent attempts to defuse this group activity and to sublimate it in a different and more productive direction. Thus, museums, parks and galleries were an attempt to forge a cultural life for the new industrial towns that reflected the bourgeois ideals of public behaviour—sober, respectable, learned and individual as opposed to the presumed working-class model of urban culture, based on more licentious and gregarious group activity.

The availability of even a limited amount of free time was thus a challenge for those in charge of urban environments, and steps had to be taken to ensure that this time was used for positive ends. Rational recreation assumed both an active middle class and a passive working class in order to be successful. In reality, this was often far from the case. Many representatives of the working class were active in raising money in the workplace to fund the establishment of public parks. This occurred at Peel Park in Salford when a total of £35,000 was raised to help to fund the park by workmen's committees in local factories and mills (*Manchester Faces and Places*, 1892, p. 54). Similarly, the London Democratic Association, a working-class movement, helped to campaign for the establishment of Victoria Park in the East End in 1838 (Hoyles, 1991).

These spaces were not always imposed from the top down, nor exclusively at the initiative of local philanthropists. The mechanisms used for their establishment were often collaborative and can rarely be judged to be entirely the product of middle-class imperialism. Such evidence of working-class agency in respect of contributions to the establishment of spaces for public recreation is often overlooked due to what Thompson has called the lack of records of the 'inarticulate majority' and the preponderance of the views and opinions of the articulate minority (Thompson, 1963, p. 59). This is also true of histories of public parks—the voice of the parks user is rarely heard as often there were no mechanisms for gathering or preserving this information.

This working-class activism has often been overlooked by historians in favour of a leisure history that privileges the role of the local civic and social elite and that defines the working classes as passive recipients of a leisure produced in its entirety by the middle classes. The history of the establishment and development of public parks presents a more complex picture of class relationships in the urban landscape. Parks, like the other sites for rational recreations, were vehicles that could be driven in varying directions. Middle-class reformers thought and hoped that they were shaping a cultural life for the working classes, but it is understandable in the case of parks that there was ample enthusiasm among the working class in a crowded, polluted city for more open space. How that open space was used and by whom was vigorously contested at times.

Attempts to regulate behaviour in public parks and to develop general guidelines with respect to their usage have also been interpreted as a form of social control. In reality, these bye-laws had a more pragmatic purpose—many people who were challenged about their behaviour in the park (e.g., picking flowers) responded by saying that they were not aware that this behaviour was prohibited. Bye-laws were often enacted in response to these actions, rather than the other way around. Attempts to delineate clearly what was and was not permitted were simply practical responses to the ways in which park users already behaved. The deployment of pithy rhetoric, such as 'parks for the people', widely used in the local press and often co-opted by local civic representatives, often conveyed the impression that there were little or no restrictions on the behaviour of the general public in any municipal park.

Bye-laws were usually prominently displayed on noticeboards around the park or near to the main entrance. These can be regarded as a form of civic socialisation—a mechanism for making the public more respectful of public property and the environment. The variety of the number and types of bye-laws also provides an interesting insight into the management styles of the civic authorities, and the differing emphases of the bye-laws display the amount and type of regulation preferred. These bye-laws were enforced either by the park-keepers (some of whom had powers of arrest) or by the local police force. Police patrols had to be paid for where they were used (£65 per annum, per constable in Birkenhead) so park-keepers were usually regarded as more cost-effective as they could perform other duties as well (LA, Parks and Gardens Committee minutes, 352/MIN/PAR/1/28, pp. 544–545). The only drawback of the park-keepers was a lack of public respect—the public did 'not heed them as they would a uniformed man' (LA, Parks and Gardens Committee minutes, 352/MIN/PAR/1/30, pp. 191–192).

More recently, the emphasis on the importance of public parks as spaces for social control and middle-class moral imperialism has begun to be challenged by historians such as Stedman-Jones (1977). He argued that 'there is no political or ideological institution which could not in some way be interpreted as an agency of social control' and thus the concept had exceeded its usefulness (p. 164). Joyce (2003) has suggested that parks were not 'elaborate disciplinary machines' at all but places that offered looser boundaries than heretofore believed (Joyce, 2003, p. 222). Indeed, the ambiguity of parks as spaces is an idea that this study will return to repeatedly, particularly in Chapters 2 and 3. More recently, Elborough has advanced the idea that parks were not founded, formed or made but 'invented' (Elborough, 2016, p. 2), again implying some agency on the part of the users. While there is little doubt that some parks were originally envisaged as places where the working classes could be exposed to 'improving' behaviour (however that was defined), it is equally clear that parks were highly contested spaces in which different sorts of normative behaviours were pursued and new kinds of urban public mores were emerging. Public parks provide an ideal opportunity to examine the emergence and mechanisms of development of these kinds of behaviours and how and where they manifested themselves.

Public Leisure and Civic Culture

The establishment of a public park enabled civic authorities to consolidate their local power structures and to visibly demonstrate their influence. While some cities, notably Cardiff, relied on land donated by local aristocrats, such as the Marquis of Bute, for their parks, others had the benefit of energetic and committed philanthropists or, in some cases, municipal representatives themselves. The mayor of Leeds, John Barran, created the impetus for the city council to buy the 775 acres of Roundhay Park for the city in 1872. The city was prevented from buying any single item worth more than £50,000, but Barran used his own wealth to purchase Roundhay and then ensured its transfer to the city via the Leeds Improvement Act, 1872 (Burt, 2000).

The official opening of a new public park was an ideal opportunity to signal publicly the benevolence and foresight of the members and to boost the image of the city to itself, its citizens and beyond. This process has been referred to as 'the invention of tradition' and was a significant component of civic ritual in the nineteenth and twentieth centuries (Hobsbawm, 1983, p. 12). Such events were a symbol of prosperity, pomp and civic pride and frequently were undertaken with the intention of promoting municipal competitiveness, especially with neighbouring localities.

The official opening ceremonies were significant municipal events, often involving parades of civic and other local dignitaries from the town hall to the new park, speeches and the presentation of a ceremonial gold key to mark the opening itself. The tone of these events was invariably self-congratulatory and acutely conscious of the symbolic significance of the occasion. Sometimes, the opening was performed by a local aristocrat (if one existed), but more often by a senior civic official, such as the mayor or chairman of the Parks Committee. Every advantage was taken to praise the foresightedness of the committee in purchasing and laying out the park and in exhorting the citizens to make good use of it.

The opening ceremony also provided an opportunity for the civic elites to develop a discourse of direct address to the citizenry that developed the idea of a city, public leisure and the role of the citizen in this. 'Citizen' was often an ambiguous term, which could be useful when being deployed for civic purposes. Citizens were active, healthy and civic-minded and 'formed the building blocks of a well-ordered society' (Harris, 1993, p. 250). These speeches tended to remind the listeners of the hard work of the civic representatives on their behalf, reinforcing the strength of the local franchise and the key role of the local government in improving the lives of all urban dwellers. There were also some rather pompous comments on the process of the acquisition or purchase of the parkland and the evidence this provided of the commitment of the civic body to the future of the city.

Members of the royal family opened some of the flagship parks, such as Liverpool's Sefton Park. Prince Arthur, son of Queen Victoria, presided in 1872 and the procession from the town hall to the park comprised 77 carriages (Twist, 2000). Not all ceremonies were so lavish as in Liverpool—possibly because the city had come later to public parks than its neighbour, Manchester. In any case, the occasion and the speeches delivered were covered in detail by the local press, ensuring the dissemination of the event to those who were not present and plenty of good publicity (Gunn, 2000).

Occasionally, a more muted tone was struck as at the opening of Horton Park in Bradford in 1878. The mayor, after performing the ceremony, took the opportunity to remind those gathered of the many difficulties encountered by Bradford Corporation when acquiring the park—multiple ownerships and complex trust arrangements followed by the arrival of a railway line in the district. These comments were augmented by the observation that parks might contribute not so much to public health but to the lessening of crimes by 'drawing people away from the allurements of the gin palace' (*The Leeds Times*, 1878).

These events tended to underline the social significance of the new urban elites in a very public way and contributed to the proto-aristocratic inclinations of many town and city council members. This was reinforced by other civic events and their ritual nature, such as the opening of new libraries and museums and funerals of former civic leaders, many of which attracted large crowds and the kinds of ceremonial distinctions usually reserved for aristocratic funerals. The absence of a permanently resident aristocracy from many British cities by the end of the nineteenth century left a social void that could be usefully filled by civic leaders and other local notables who had pretensions to behave in ways that were publicly visible and that attempted to constitute a civic tradition that represented and contrasted with the absent nobility.

These new civic elites or 'urban aristocracy' in Briggs's words were motivated by a general concern about what were regarded as the dissolute lives of the working classes, which revolved around drinking and gambling (Briggs, 1968, p. 94). This concern had given rise to the development of libraries, art galleries and museums, in part to provide a series of spaces in which the working classes could learn to better themselves.

Such philanthropy was supposed to work as a means of improving society as a whole. The public houses often used by the working classes had themselves contained libraries and sporting facilities (Bailey, 1978). The replication of some of these amusements in the newer environment of the library and art gallery was aimed at attracting the working classes away from the public houses and providing a more supervised learning environment.

These occasions and the spectacle they provided were enthusiastically reported on in detail by the local press. The further circulation of these descriptions of the civic rituals that pertained to these events consolidated the impression in the ordinary citizen that their public representatives were actively contributing to the common good and emphasised the importance of the public park as a place of civic ritual and spectacle. Many local newspaper accounts accepted the civic doctrine being promoted and reinforced both the invention and extension of the civic tradition that they represented. There was much detailed direct quotation from the speeches, which not only conveyed a flavour of the event to those who did not attend but also reproduced effectively and without criticism the themes of continuity and municipal benevolence. This implicated the local press firmly and uncritically within the doctrines of civic and self-improvement and furthered the cause of rational recreation well into the twentieth century.

Ideas about improvement and the common good were fundamental to much of Victorian thinking about life in the city. Parks in many cities and towns were developed not from the relevant parks committee per se but under the auspices of improvement. Many parks committees began as improvement committees and were gradually devolved, as were bath and washhouses, libraries, museums and art galleries. Like the common good, improvement carried within it connotations about Victorian life and values. Noble has defined the common good as bringing utility to as many members of a community as possible, while acknowledging that, like improvement, the word was 'semantically unstable' and often contested, especially with regard to the cost of such provision (2016, p. 41).

The obverse of improvement was often economy (Fraser, 1979). Grandiose plans for the civic realm had to be financed and this frequently proved to be controversial. As Fraser has suggested, opposition to public expenditure was more about questioning what the money would be spent on and not the expenditure itself. In the case of public parks provision, the debate often focused on the location of the park, who the beneficiaries would be and who would inevitably lose out. Improvement was, therefore, a political matter. It often had Liberal underpinnings and, as Chandler has argued, 'many improvement initiatives coalesced around a Whig-Liberal alliance supporting an innovative municipal corporation as opposed to an older oligarchy of predominately Tory interests seeking to maintain the status quo' (2007, p. 71). However, care must also be exercised not to generalise too much from this and to develop a nuanced approach to the functioning of local government in the period under consideration. Opposition to improvement of the public realm, such as parks and museums, was often difficult in practice, and it was by no means always the case that such initiatives were always proposed by Liberals and opposed by Tories. These aspects will be explored further in Chapters 4 and 5.

As the nineteenth century advanced into the twentieth, two groups became the focus of attention in respect of the provision of leisure facilities—women and children. The Victorians had been cautious about the need for and respectability of leisure activities for women. In many cases, ideas about improvement and self-help were underpinned by universal access, but this was not always so in respect of women's leisure (Rodrick, 2004). From their inception, public parks had been popular with women. Contemporary photographs show substantial numbers of women and children in parks, especially during the week. Women were often employed in the tearooms as catering staff and as lodge keepers. They were also employed by city councils to supervise the play of young children during the summer months, thus serving an important didactic function.

The present study provides a rare opportunity to examine the working lives of those employed in public parks, detailed in Chapter 4. These parks departments were often substantial employers, yet we know little about the daily working lives of these people and how they viewed their jobs. Much of the work was hard, manual labour but with some skilled and semi-skilled professionals who worked as gardeners, park-keepers, park superintendents and chief superintendents. The operation of these hierarchies and their working relationships with both those who used the parks and the wider municipal bodies shed light on the development of professional training and working practices in public parks, but also on the growth of municipal bureaucracy during the twentieth century.

The expansion of public leisure demanded an increasingly professional approach to park management and the development of mechanisms for employing, training, developing and disciplining staff. An expansion of the numbers of large parks and smaller recreation grounds during the twentieth century placed greater demands on the role of chief parks superintendent also. This individual was responsible for the development and management of all parks and recreation grounds (and, in some cases, municipal burial grounds or public bathhouses) within a municipal area and occupied an influential position between the parks committee and its employees. He had to be not just an experienced horticulturalist but also a skilled people manager and administrator. Prominent among this profession during the period of this study were the Pettigrew brothers—Andrew and William, chief parks superintendents of Cardiff (1915–1936) and Manchester (1915–1932) respectively (William was also chief parks superintendent in Cardiff (1891–1915), before his brother). Well known and respected in their field, their combined influence was substantial. They established the professional basis for parks administration in twentieth-century Britain and their legacy to the tradition of public recreation remains observable today.

Schenker has suggested that the nineteenth-century park presented its users with an enigmatic experience that was often culturally ambiguous (Schenker, 2009). He argues that this was intensified by a public that did not know how to respond to these spaces. While acknowledging that parks designers attempted to orchestrate the public's responses, it must be accepted that much of the public response could not be anticipated. It is true that the public made their own use and interpretation of these parks. They responded in ways that were often unexpected and that were sometimes negative. They committed crimes and acts of vandalism, they engaged in sexual activity, drank and gambled, and killed other people and themselves. These forms of behaviour revealed public parks to be places where conflicting ideas about public behaviour, leisure and citizenship all emerged and were debated.

A Note on Sources

One of the pleasure and challenges for the historian of public parks is both the amount and diversity of sources available. Primary sources, such as City and Town Council Parks Committee minutes, form the backbone of this study and can offer a valuable insight into the decision-making processes about park-making and development. Often, these decisions were taken alongside other municipal services, such as cemeteries, libraries, art galleries and baths. There was a particular affinity between parks and cemeteries due to their contemplative atmosphere and so we often find them being considered together by a municipality.

Documents, records and reports by the parks superintendent, where they exist, also provide a valuable source of studying the nexus with the relevant Parks Committee. While a parks superintendent reported to this committee, they also made recommendations and policy decisions that were directly adopted by the committee. Their expertise carried weight and they were frequently called upon to dispense advice within that city government and occasionally to other cities and local authorities. Access to their thinking and priorities provides an insight into the development and importance of the role played by the parks superintendent in deciding the direction of parks policy and also in providing a link between the municipal representatives and those actually employed in the parks.

We know tantalisingly little about those employees and those who visited and used the parks in their early years. Historians have lately begun to gather oral testimony of people's more recent memories of parks visits, but those of many of the earliest visitors have gone forever. Contemporary accounts can be found in local newspapers and periodicals, in diaries and autobiographies and in pamphlets and other tracts, but these are sporadic at best. Newspapers in particular are a significant source of debates about leisure more generally and were an important arena in which emerging ideas could be aired and tested.

Sources of information about the events and activities in public parks are more various. The development of these spaces as multifunctional arenas emphasises the need to survey and include a more complex variety of information sources. These include records of sporting events, political meetings and demonstrations, crime data, wartime and military records and those produced by the many bodies involved in the emerging health, fitness and environmental movements of the early twentieth century. The variety of voices that these afford offers an important adjunct to the official records of the municipal bodies and allows us to determine to some extent the experiences of the ordinary parks visitor.

The choice of cities to be covered in this study (Manchester, Salford, Liverpool, Leeds, Preston, Hull and Cardiff) was dictated by geographical and archival factors, principally. The availability of the records of the municipalities involved, along with a decision to examine those urban areas that had suffered most from industrialisation and who, arguably, had most to gain from the provision of open space for recreation, was also a factor. Cardiff was included principally due to the impact of the Pettigrew brothers, Andrew and William Wallace, on the development of parks both in Cardiff itself and in Manchester. The decision to end the study at the year 1939 was determined by the obvious outbreak of World War II and the sheer amount of information that the study had yielded by that date. Undoubtedly, more studies are needed, of varying lengths and with differing geographical foci. The trajectory of urban parks developments is uneven and contested, and any decisions about these studies will result in a necessarily partial picture.

Conclusion

In her influential 1961 work *The Death and Life of the Great American City*, Jane Jacobs described parks as 'volatile' spaces and argued that too much was expected of such places in terms of improving quality of life and access to open, green space (Jacobs, 1964, p. 99). While Jacobs was mostly describing what she called 'neighbourhood parks' in American city centres, much of her argument could also be applied to larger public parks on the edges of urban areas. Written at a time when many public parks were in the middle of a long period of sustained decline, this view seems understandable. We should be wary of seeing public parks as a universal good or a panacea for all social ills. They certainly had the *potential* to contribute to the improvement of public health and recreation, but this should not be taken for granted. Such assumptions about the ubiquitous nature of public open space underpin much of the writing on urban parks and risk not just a tendency to generalise about their history and social impact but also a failure to appreciate the opportunities to learn from their shortcomings.

It would be a mistake to approach the study of the urban park as an evolutionary process from its origins to a late nineteenth-century 'golden age' to a period of gradual decline in the twentieth century. Such assumptions skew our sense of historical process and imbue parks with a kind of inevitable trajectory, often outside of their own control. Urban parks meant different things in different locations; they developed at varying speeds, based very often on localised and specific factors. It is this difference that makes their study both necessary and challenging.

The themes identified in this chapter—public health, citizenship, normative behaviour and public leisure—all contribute to our understanding of these spaces and how they worked in various urban settings and at various times. They also have a significant contemporary resonance—many urban parks remain sites of contestation and conflict. They demand constant upkeep and renewal; they mean different things to different communities. They develop around them communities of belonging that change over time. These communities are also not in themselves consistent and consensual. They are inclusive and exclusive; they exist to maintain boundaries and to contain us but also to offer a challenge to those controls.

While the original nineteenth-century parks were intended to provide a rural contrast to the industrial overcrowded city, by the twentieth century, they were being incorporated into the wider urban landscape by virtue of newer town planning schemes influenced by the garden city movement. In conjunction with other aspects of urban greenery, such as street trees and parkways, they formed part of a broader consideration of the role of open green space in the planned city.

Parks engage people, they cause and stimulate dissent and they can (potentially) celebrate both difference and similarities. In trying to understand how they function, we must consider a variety of factors and agents, both internal and external, local, regional and national. The boom in the development of public parks in so many British towns and cities provides an insight into the movement of social, economic and cultural power from the private, aristocratic estates and country houses into town halls, art galleries, parks, museums and libraries. As such, this is the story of the awakening of municipal government and the emergence, consolidation and eventual diminution of civic leadership at local level as expressed through the provision of public space for recreation.

References

Bailey, P. (1978). *Leisure and Class in Victorian England*. London: Routledge.

Borsay, P. (2010). Walking on the Urban Edge: Peripheral Space and Recreation in English and Welsh Towns c. 1700–1900. Paper presented at European Association for Urban History Conference, Ghent, Belgium.

Briggs, A. (1968). *Victorian Cities*. London: Penguin.

Burt, S. (2000). *An Illustrated History of Roundhay Park*. Leeds: S. Burt.

Carter, J. (2011). Middle Class Aspirations in the 'People's Park': Birkenhead and a Radical Working Class Ideology. Unpublished paper.

Chandler, J. (2007). *Explaining Local Government: Local Government in Britain Since 1800*. Manchester: Manchester University Press.

Cobbett, W. (1812). *Cobbett's Parliamentary Debates* (Volume 11). London: Longman.

Conway, H. (1991). *People's Parks: The Design and Development of Victorian Parks in Britain*. Cambridge: Cambridge University Press.

Elborough, T. (2016). *A Walk in the Park: The Life and Times of a People's Institution*. London: Jonathan Cape.

Fraser, D. (1979). *Urban Politics in Victorian England: The Structure of Politics in Victorian Cities*. Basingstoke: Macmillan.

Garrard, J. (1977). *Leaders and Politics in Nineteenth Century Salford: A Historical Analysis of Urban Political Power*. Salford: University of Salford Department of Politics Occasional Papers.

Gunn, S. (2000). Ritual and Civic Culture in the English Industrial City c. 1835–1914. In R. J. Morris and R. H. Trainor (Eds.). *Urban Governance: Britain and Beyond Since 1750* (pp. 226–241). Aldershot: Ashgate.

Harris, J. (1993). *Private Lives, Public Spirit: A Social History of Britain*. Oxford: Oxford University Press.

Hill, K. (2001). 'Roughs of Both Sexes': The Working Class in Victorian Museums and Art Galleries. In S. Gunn and R. Morris (Eds.). *Identities in Space: Contested Terrains in the Western City Since 1850* (pp. 190–103). Aldershot: Ashgate.

Hobsbawm, E. J. (1983). Introduction: Inventing Traditions. In E. J. Hobsbawm and T. Ranger (Eds.). *The Invention of Tradition* (pp. 1–14). Cambridge: Cambridge University Press.

Hoyles, M. (1991). *The Story of Gardening*. London: Journeyman Press.

Hunt, T. (2004). *Building Jerusalem: The Rise and Fall of the Victorian City*. London: Phoenix.

Jacobs, J. (1964). *The Death and Life of Great American Cities*. London: Penguin.

Joyce, P. (2003). *The Rule of Freedom: Liberalism and the Modern City*. London and New York: Verso.

Kay, J. P. (1832). *The Moral and Physical Condition of the Working Classes Employed in the Cotton Manufacture in Manchester*. London: James Ridgway.

Kidd, A. (2006). *Manchester* (4th Ed.). Lancaster: Carnegie Publishing.

Layton-Jones, K. and Lee, R. (2008). *Places of Health and Amusement: Liverpool's Historic Parks and Gardens*. Swindon: English Heritage.

Levin, M. (1998). The Condition of England Question: Carlyle, Mill, Engels. Basingstoke: Macmillan Press.

Liverpool Archives, Liverpool, Parks and Gardens Committee Minute Books, 352/MIN/PAR/1/.

Manchester Borough Police Act, 1844, 7 & 8 Vict., c. 40.

Manchester City Council (1880, 1 May). *Manchester Courier*, p. 19.

Mosley, S. (2001). *The Chimney of the World: A History of Smoke Pollution in Victorian and Edwardian Manchester*. London and New York: Routledge.

Noble, M. (2016). The Common Good and Borough Reform: Leicester c.1820–1850. *Midland History*, 41(1), 37–56.

Meller, H. (1976). *Leisure and the Changing City, 1870–1914*. London: Routledge and Kegan Paul.

The Opening of Horton Park, Bradford (1878, 1 June). *The Leeds Times*, p. 2.

O'Reilly, C. (2013). From 'the People' to 'the Citizen': The Emergence of the Edwardian Municipal Public Park in Manchester, 1902–1912. *Urban History*, 40(1), 136–155.

Peel Park (1892, 11 January). *Manchester Faces and Places*, 3(4), 54–55.

Pettigrew, A. A. (1926–1932). *The Public Parks and Recreation Grounds of Cardiff*, 6 Volumes. Unpublished.

Pettigrew, W. (1937). *Municipal Parks: Layout, Management and Administration*. London: Journal of Park Administration.

Pollard, W. (1882). *A Handbook and Guide to Preston*. Preston: H. Oakey.

Poovey, M. (1995). *Making a Social Body: British Cultural Formation 1830–1864*. Chicago and London: University of Chicago Press.

Public Health Act, 1925, 15 & 16, Geo V, c.71.

Redford, A. (1939). *The History of Local Government in Manchester*. London: Longmans Green & Co.

Rodrick, A. (2004). *Self-Help and Civic Culture: Citizenship in Victorian Birmingham*. Aldershot: Ashgate.

Salford Town Council (1897, 3 March). *Manchester Evening News*, p. 2.

Schenker, H. M. (2009). *Melodramatic Landscapes: Urban Parks in the Nineteenth Century*. Charlottesville and London: University of Virginia Press.

Simon, E. (1926). *A City Council From Within*. London: Longmans Green & Co.

Slaney, R. (1860, 15 May). Recreation Grounds Debate. Hansard, Volume 158, Column 1288. Available from: https://hansard.parliament.uk/Commons/1860-05-15/debates/3b9a72b1-2247-4fae-9f59-f7e4cdb79228/Resolution?highlight=1833%20select%20committee%20public%20walks#contribution-b89500e1-3e68-4213-b9a9-bb8699ccc377

Stedman-Jones, G. (1977). Class Expression Versus Social Control? A Critique of Recent Trends in the Social History of 'Leisure'. *History Workshop Journal*, 4(1), 162–170.

Stevenson, J. (1984). *British Society 1914–1945*. London: Penguin.

Swift, R. (2001). Thomas Carlyle, 'Chartism', and the Irish in Early Victorian England. *Victorian Literature and Culture*, 29(1), 67–83.

Taigel, A. and Williamson, T. (1993). *Parks and Gardens*. London: BT Batsford.

Taylor, D. (1999). Central Park as a Model for Social Control: Urban Parks, Social Class and Leisure Behaviour in Nineteenth Century America. *Journal of Leisure Research*, 31(4), 420–477.

Thompson, E. P. (1963). *The Making of the English Working Class*. London: Penguin.

Twist, C. (2000). *A History of the Liverpool Parks*. Southport: Hobby Publications.

Williamson, T. (1998). *Polite Landscapes: Gardens and Society in Eighteenth Century England*. Stroud: Sutton Publishing.

Wyborn, T. (1995). Parks for the People: The Development of Public Parks in Manchester. *Manchester Region History Review*, 9, 3–14.

Zweiniger-Bargielowska, I. (2010). *Managing the Body: Beauty, Health and Fitness in Britain 1880–1939*. Oxford: Oxford University Press.

2 Parks in the Urban Environment

Introduction

Urban parks were originally designed to offer a haven from heavily polluted and congested cities. Ironically, they quickly became so popular that they resulted in landscapes almost as overcrowded as the cities themselves. The idea of a freely available open, green space that combined areas for walking and sporting activity dominated debates about public health in the nineteenth century. A letter writer to the *Leeds Mercury* in 1869 wrote of the need in that city for 'a boasted panacea for all the ills which afflict our pent-up people' (*The Leeds Mercury*, 1869). The Victorians conceived of public parks as places in which the moral character of the working classes could be improved, but this idea rapidly evolved into envisaging parks as places to train the urban citizen to fulfil his or her role in the city and beyond. A new emphasis on securing the empire resulted in the need for a population that was physically fit and healthy enough to fight for its future.

Parks were thus reconfigured as spaces in which the visitor could learn the vital citizenship skills deemed necessary through contact with such organisations as the Boy Scouts. The early twentieth-century park moved away from the Victorian idea of 'parks for the people' in which the resolution of social problems could be achieved by the mingling of all social classes into an emphasis on active citizenship, where social responsibility was equally shared by all. Urban parks did not develop in isolation from other elements of the city. The existence of urban green space inspired the garden city movement of the early twentieth century, and innovations like the National Playing Fields Association (1925) were founded as a result of the popularity of parks as sites of sport and recreation.

Urban planners have often approached the design and planning of urban space as an attempt to amalgamate the virtues of the rural environment in an urban location—*rus in urbe*, so called. On many occasions, however, the vicissitudes of city life derailed this effort. Public parks, once envisaged as green lungs for the city, ended up as crowded and densely attended as any urban space. Many citizens (especially those with the necessary means) surrendered to the desire to escape the city for the countryside or even the suburb (Briggs, 1963). This flight was facilitated and accelerated by the arrival of the railway in the early nineteenth century. The dominance of the motor car during the twentieth century resulted in an increasingly difficult city environment in terms of the wish to establish open, public spaces, and the attempt to 'green' the city ended in stalemate.

Rational Recreation: Parks in the Victorian Period

The genesis of the need for urban parks was the 1833 Select Committee on Public Walks, which was established to report on the provision of recreational facilities in overcrowded and polluted British cities (Select Committee on Public Walks, 1833). The appellation 'public walk' was generally applied to these early parks, emphasising their genteel purpose and emphasis on creating long, winding walkways bordered by planting and lawns. There, the city dweller could contemplate the horticulture while strolling in a peaceful and quasi-rural setting. The report remarked that

> with a rapidly increasing population, lodged for the most part in narrow courts and confined streets, the means of occasional exercise and recreation in the fresh air are everyday lessened, as inclosures take place and buildings spread themselves on every side.
> (Select Committee on Public Walks, 1833, p. 5)

It singled out cities such as Manchester as particularly in need of these spaces due to the temptations of alternative pursuits, such as drinking and gambling. Inscribed within the report was a series of long-held assumptions about the urban life of the working classes. A fundamental connection was made between physical and moral health in the early park. This came to be known as 'rational recreation' and was aimed, initially, at the working classes, who, it was believed, needed to be saved from the twin evils of drinking and gambling. Explicit contrasts were drawn between life in the overcrowded city and the romantic possibilities offered by a stroll in 'an immense expanse of greensward, grateful and refreshing to the eyes that have poured over ledgers and accounts all day' (Sullivan, 1915, p. 74). It is interesting to note that the park user envisaged here is a professional and not a member of the urban poor. Thus, the ideal of 'parks for the people' was already waning.

The pleasure to be derived from walking was implicit, as it was in the 1833 Report, and was presumed to be an attractive alternative to other, less acceptable pursuits. Walton has demonstrated that the working classes merely incorporated visits to these institutions into their lives without necessarily giving up what was perceived to be the working-class penchant for gambling and drinking (Walton, 1987). Thus, public parks co-existed with other, less rational forms of recreation, rather than acting as a substitute.

A further impetus for the development of urban parks was provided by the 1848 Public Health Act, which included (voluntary) empowerment for Local Boards of Health to provide public walks or pleasure grounds in cities (Public Health Act, 1848, Section 74). The pleasures of the public park, it was hoped, would encourage the working classes to better themselves and their health, and many early parks had facilities such as art galleries and museums to add an educational dimension to their physical health properties. Many British cities had developed significant public spaces, ranging from public squares to shopping arcades and market-places, by this time.

Urban streets were also often regarded as open spaces in the Victorian period, especially in working-class areas, where access to alternative spaces was restricted. It is important to note that public spaces and their uses have always been contested. The struggle over access to public space and for whom it is intended has been a long-established part of city life. City squares, halls and meeting places have, in part, symbolised the democratic nature of urban life—freely open to all who can or wish to access them but, at times, regulated, patrolled and highly managed. This reveals a public space that is deeply politicised and that has been used to express provocative and controversial views that challenge the status quo—from the Chartists to the suffragettes to the British Union of Fascists. Thus, urban public space can fulfil dynamic functions and contain people's movements that have the power to alter history.

Parks emerged as a significant element in this attempt at social engineering, along with libraries and museums. Rational recreation has provided a popular explanation for the development of municipal parks during the Victorian period, but it has also limited our understanding of the role of the urban park, with an overemphasis on usage and elements of park design and neglecting park management decisions and municipal priorities.

Rational recreation grew out of an attempt to provide role models for public behaviour and to encourage the adoption of the values of a new urban middle class, which considered itself both culturally and morally superior (Wyborn, 1995). Thus, parks are often viewed as representing middle-class moral imperialism or social control. Conway has offered a more subtle view of rational recreation and the problematising of working-class leisure (Conway, 1991). She pointed out that there are two possible interpretations of rational recreation—one that accords with Wyborn and one that implies a more understated approach (Conway, 1991). The latter sees rational recreation as an attempt to enhance and widen the cultural experiences of the working classes more indirectly. There were certainly attempts to educate park visitors in art appreciation and to supervise the play of children, but these were basic measures of instruction and self-improvement that were commonplace in the Victorian period. This happened in environments other than public parks, such as churches and schools; its effect is difficult to assess and should not be overemphasised for this reason.

Public parks were associated with ideas about public health and recreation but also with providing 'green lungs' for overcrowded and polluted cities. The early parks were conceived of as providing an alternative to the city landscape and valued for their open spaces and connections with the rural (Sullivan, 1915). In practice, this was rarely the case. So popular did the early parks prove that they were often as overcrowded as the city streets, especially at the weekends and on public holidays. The park experience was often compared to spending time in the countryside, and this was an especially popular parallel in the pages of the local newspapers. Many commentators in the local press took a rather romanticised view of the contribution of public parks to their towns and cities. 'Quiz' writing in the *Hull Packet and East Riding Times* in 1876 juxtaposed the rural nature of the park with the urban attributes of its visitors— 'pleasantly blending with the sweet essence from a hundred flowers is the fragrant aroma of the soothing cigar'—and advocated 'throwing oneself at full length upon the velvet grass' (*Hull Packet and East Riding Times*, 1876). A similarly florid account of the Manchester parks from 1902 referred to 'how the hand of man can impart beauty to nature' (*Manchester Courier*, 1902).

It should also be noted that, at municipal level, parks were often administered along with cemeteries, which, in the Victorian period, were also highly landscaped public spaces designed for quiet contemplation. Liverpool was one of the first British cities to establish a cemetery at St James's in 1825 (Curl, 1975). The landscape designer, John Claudius Loudon, believed that, aside from the practical function of cemeteries, they should serve as a 'school of instruction in architecture, sculpture, landscape gardening, arboriculture . . . and botany' (Curl, 1983, p. 141). Penny has mooted that such cemeteries deserve to be considered as the successors to the eighteenth-century private gardens (Penny, 1974). Liverpool's Grant Gardens was built on the site of the necropolis at West Derby Road, emphasising the unity of cemetery with public park. The necropolis opened in 1825 and closed in 1898, with the new park opening in 1914.

There were often practical barriers to the use of a former cemetery as a park. All Saint's cemetery in Manchester was initially suggested for use as a public park in February 1910. An estimate of £3,000 to convert the 8,511 square yards of space to a public park, including lowering and covering all of the gravestones and listing each one and its inscription, was regarded as excessive (MA, Parks and Cemeteries Committee minutes, Volume 29, p. 110). The city surveyor prepared three alternative plans costed between £2,300 and £2,500 in July 1910 (MA, Parks and Cemeteries Committee minutes, Volume 29, p. 179).

A report of the plans in a newspaper, the *Manchester Courier*, prompted some letters of complaint, principally from Canon Joseph Nunn, rector of St Thomas's, Ardwick (MA, Parks and Cemeteries Committee minutes, Volume 30, p. 18). Nunn expressed concern about the families of those buried at the site and about the location being 'situated in the immediate neighbourhood of many common lodging houses and places of amusement' (MA, Parks and Cemeteries Committee minutes, Volume 30, p. 72). Nunn's fears were contradicted by a report from the Medical Office of Health, which advised that the provision of the park in the area of All Saint's would 'be a great boon to mothers with young children' (MA, Parks and Cemeteries Committee minutes, Volume 30, p. 84). This reflects a trend in the direct provision of leisure amenities suitable for women and children in the early decades of the twentieth century. The former cemetery was eventually opened as a public park in 1935. In his opening address, the chairman of the Parks and Cemeteries Committee, Alderman Will Melland, praised the provision as evidence of the fact that the city had 'wakened up considerably . . . to the need for open spaces for our young children' (*Manchester Guardian*, 1935).

Many British cities—Leeds, Liverpool, Manchester, Cardiff, Glasgow—enthusiastically acted on the recommendations of the 1833 Select Committee and began to acquire land for parks. Some of the most influential landscape designers in Britain worked on public parks—John Claudius Loudon and Joseph Paxton. Loudon designed Derby Arboretum in 1840. Paxton had worked for the Duke of Devonshire at Chatsworth and designed Birkenhead Park (1845) and Kelvingrove Park in Glasgow (1854). Most of these parks were strikingly similar in terms of their design—long, winding walkways, bordered by planting schemes and containing gymnasia with basic recreation equipment. Many of the original park designers remained involved in the parks as they developed. Edouard Andre, one of the original designers of Liverpool's Sefton Park, visited and inspected the park in 1904 with a view to modifying the design for the new century. His proposals were modest, possibly for budgetary reasons, and he confined his recommendations to the thinning of some of the trees and the alteration of some pathways.

The undulating walkways that were such a characteristic of the larger public parks were not universally appealing, however. In 1907, the secretary of the Manchester Field Naturalists and Archaeologists Society wrote to the chairman of the Parks and Cemeteries Committee to complain of 'the sense of dreariness which impresses the mind when traversing the long, naked walks' at Heaton Park (MA, Parks and Cemeteries Committee minutes, Volume 26, p. 238). Many public parks were gloomy places during the winter months, when the trees lost their leaves and visitor numbers declined. This comment is also an indication that, from a design perspective at least, the idea of a controlled and carefully designed landscape as advocated by Repton and Loudon was losing its appeal during the early twentieth century.

It is equally likely that public parks were evolving their own design aesthetic as opposed to borrowing from already-existing aspects of private, aristocratic parks. There were some tensions among parks superintendents about this issue. In 1928, the Royal Horticultural Society held its first International Exhibition of Garden Design in London, which was attended by many of those responsible for the design and administration of public parks in Britain. Manchester's parks superintendent, William Wallace Pettigrew, gave a lecture at the event on the subject of public park design. He disparaged the conventional design features of the early public park, such as formal gardens, curved walkways and carpet bedding, and promoted more modern aspects, such as straight lines. The chief superintendent of Liverpool's parks, J. J. Guttridge, in his report on the event, commented that there was no time allocated to debate the subject with Pettigrew, as he (Guttridge) believed that the more traditional design elements were more appealing to the public (LA, Parks and Gardens Committee minutes, Volume 32, pp. 290–291).

F. E. Seton, writing about the exhibition in the *Spectator*, echoed Pettigrew's dissatisfaction with formality in garden design:

> The design which first asserts itself by constant repetition is the formal garden and one feels that its curse is monotony. Always it takes the form of a long oblong laid out in the same design, a straight central walk swelling in the middle to contain some object, such as a flat lily pool. On either side a broad stretch is cut into numerous small, triangular beds filled with fussy little plants and divided by narrow zigzag paths.
>
> (*The Spectator*, 1928)

Pettigrew's approach to this issue displayed his usual pragmatism. Writing further on this idea in 1937, he argued that the use of straight lines within public parks made for a more efficient use of the space and left more room for revenue-generating activities, such as tennis and bowls. 'The designer who has a predilection for beautiful curving roads and walks, at one time so dear to the heart of the landscape gardener', he wrote, 'will find it difficult to confine himself to the more utilitarian ideas necessary in modern public park administration' (Pettigrew, 1937, p. 6). He did suggest that the public taste for formal gardens could be accommodated if space permitted and in a separate part of the park, away from the sporting attractions (Pettigrew, 1937). Thus, public opinion could be somewhat assuaged, without detracting from the parks' ability to generate income.

Some city councils were concerned about the state of public opinion about the provision of open space and how it was to be financed. What

Laski has termed the 'angry eye of the ratepayer' did not restrict the entrepreneurial flourishes of authorities such as Manchester City Council, whose commercial success was well established by the mid-Victorian period (Laski, 1936, p. 107). Urban parks would require long-term investment to be fully functioning, and this would often prove to be controversial, however universal the need for public recreational space. In Cardiff, the Local Board of Health debated the need for public approval for the acquisition of land for public parks. One member, Mr Thomas, protested: 'The public have not expressed their opinion about it'—that is, upon the desirability of providing recreation grounds—'I have not heard it out of this room' (Pettigrew, Volume 1, 1926, p. 49). The difficulty of assessing the strength of public feeling on the issue resulted in an overly cautious attitude on the part of many local authorities. Garrard (1983) has outlined how a proposal in Rochdale to make the Glebelands into a public park was concluded only after seven years, partly due to anxiety about public opinion on the issue.

Where the principle of providing public parks was accepted, there remained some timidity among some local representatives about how best to determine the opinion of the ratepayers on the subject. The Lord Mayor of Manchester, John Foulkes Roberts, suggested that the local press could be a useful tool for assessing public opinion on the subject. He asked, '[H]ow can a Corporation complete a large purchase of land without the certainty that the ratepayers support the proposal and how is that support to be ascertained except through the medium of the press?' (*The Manchester City News*, 1897). In general, the local press was supportive of plans to acquire and develop public parks, enthusiastically adopting the phrase 'parks for the people' as a suitable journalistic rhetorical device. Much of this comment reflected a spirit of municipal competitiveness or 'civic boosterism' that characterised this period (Stobart, 2003, p. 169).

Municipal authorities responded to this perceived favourable public opinion either by acquiring land for parks or by accepting donations from wealthy local aristocrats and philanthropists, as happened in Cardiff, many of whose parks originated from land donated by the Marquess of Bute (Pettigrew, 1929, Volume 2, p. 7). Some parks resulted from the individual visions of wealthy and socially prominent members of the local authority. In Leeds, the campaign to buy Roundhay Park for the city was led by the mayor, John Barran, in 1871 (Burt, 2000). As the city was not permitted to pay more than £50,000 for a single item, Barran bought the park himself for £139,000 and used the Leeds Improvement Bill to obtain the permission of Parliament for the purchase (Burt, 2000).

Regulation of behaviour in these open and accessible spaces also presented municipal authorities with a challenge. The 1875 Public Health Act provided for the formal regulation of public walks and for the potential removal of any person from the park who infringed the bye-laws

(Public Health Act, 1875, Section 164). The intrusion into the park of elements of urban life continued, with some municipal authorities selling off parkland for building to defray the costs of park acquisition and design. The mere presence of a public park in a city could influence the urban development around it—this was demonstrated with the construction of a series of high-class residential developments at Park Terrace and Park Gardens around Kelvingrove (or West End) Park in Glasgow during the 1850s (Maver, 1998). Here, the layout echoed the curve of the park, while the main thoroughfares led to and from the park entrance.

The design of the entrances to the parks is worthy of further comment. Many of the larger parks had multiple entrances, the significance of which was clearly delineated by the design and structure of the entranceways. In many cases, imposing gates and stone pillars signalled the principal entrance. Some entrances were kept quite plain, as at Middleton Park in Leeds (1920) (Figure 2.1). Other main entrances were marked by the presence of a substantial lodge house. These buildings were inhabited by lodge keepers, whose responsibility it was to control access to the park and to open and close the gates at the designated times (e.g., Chorlton Lodge, Alexandra Park, Manchester 1868). Bradford's Lister Park (1870), the former estate of the Lister family, had its main gates erected in 1904 to commemorate the opening of the Bradford exhibition at the park.

Most of the main entrances of public parks had carriage entrances with two secondary pedestrian entrances on either side. The most ornate gates were reserved for the main entrance, those of the secondary entrances being plainer and more functional. Salford's Victoria Arch in Peel Park was erected to commemorate the Queen's visit there in 1857 (Figure 2.2). The Grand Entrance to Birkenhead Park on the Wirral celebrates the splendour of the park within (Figure 2.3). Many public parks were surrounded either by railings or by boundary walls. These often

Figure 2.1 Entrance gates, Middleton Park, Leeds.

Source: © By kind permission of Leeds Library and Information Service (www.leodis.net).

Figure 2.2 Victoria Arch, Peel Park.

Source: © Salford Local History Library Collection.

Figure 2.3 Birkenhead Park entrance.

Source: © By permission of Historic England Archive.

predated their use as parks, especially if they had been former aristo-cratic estates. The boundaries were significant in determining the differ-ence between parkland and the surrounding urban area. They marked the point at which the park visitor moved from one type of space into another. This boundary had an important psychological dimension as well as a physical aspect. Sociologists have long argued that boundar-ies and their maintenance serve a crucial function in helping to deter-mine the limits of behaviour in any kind of social environment. These boundaries assist in measuring both acceptable and unacceptable forms of conduct and in judging what kinds of experiences belong to a particu-lar domain (Erikson, 1960). The park boundary performed this function then by signalling that this was a space in which the social norms of the city could be open to challenge or at least to variation. It indicated that this was a new type of space with new possibilities and with new kinds of behavioural standards.

The symbolic significance of the park boundary and its many variations played a role in alerting the park user to the singularity of the park environment and its distinctiveness. The boundary was intended to mark out the parkland and to inform the visitor of its rules. It was no coincidence that, in the main, the regulations were prominently displayed near to the main entrance. However, this could also have had the opposite effect. There are many instances of the attempts to inhibit or prohibit certain forms of behaviour that have the effect of actually encouraging it instead (Erikson, 1960). Public parks often attracted marginalised individuals, such as vagrants, and gave them the opportunity to meet and to replicate each other's behaviour. Thus, the park boundary both demarcated and integrated the space into the city environment.

Another significant design aspect was what Jane Jacobs has called 'centring' (Jacobs, 1964, p. 114). This refers to the climactic point of the park, often signalled to the visitor by the use of a focal point, such as an ornate bandstand, a statue of a member of the royal family (Queen Victoria, Prince Albert) or of a member of a prominent local family, or a drinking fountain. These were not always placed in the geographical centre of the park but could be used to guide the visitor around the space and to indicate the beginning or end of a main pathway. In some parks, several pathways could conjoin at this focal point and many often radiated out from this centre.

Decisions about the design and layout of these parks were often driven by the need to sell some of the land to fund the new development. The sale of parkland itself for building had begun in Birkenhead Park, when 60 acres were sold, while Liverpool City Council sold 110 acres of Sefton Park, which saw some 90 villas built by 1890 (Twist, 2000). This tactic was not always successful or desirable, however. Sales of land in Roundhay Park Leeds were slow during the 1860s, while Manchester City Council chose not to sell any parts of Heaton Park, acquired in 1901, for fear of antagonising the neighbouring borough of Prestwich, which had allowed the city to incorporate the park after its acquisition (Burt, 2000 and O'Reilly, 2009).

Where land had to be purchased to provide public parks, many city councils were keen to make a good bargain, where possible. If an estate did not prove suitable, for whatever reason, it was not proceeded with. The trustee of Prince's Park in Liverpool, Henry Yates Thompson, commented that the chairman of the Parks and Gardens Committee 'seems to be a good bargainer' when negotiating the transfer of the park to Liverpool Corporation in 1909 (LA, Parks and Gardens Committee minutes, Volume 26, p. 15). Manchester's Parks and Cemeteries Committee bargained repeatedly for Heaton Park throughout the late 1890s before finally acquiring it for £230,000 in 1901. The city had failed to purchase the much larger (and less suitable) Trafford Park in 1896. Similarly, Liverpool rejected the opportunity to buy Thingwall House estate in 1905 due to the number of public footpaths and rights of way on the property (LA, Parks and Gardens Committee minutes, Volume 24, p. 69). The local press was frequently blamed for exciting public opinion about the potential acquisition of a new park and for hampering negotiations with their enthusiasm and thereby driving up the price asked for by the owner. Evidence of this is difficult to find, but it remained a fervent belief of many councillors involved in thwarted negotiations for parkland and eager to place the blame for their failures elsewhere.

Many poorer areas continued to press the case for the provision of a public park in their environs. The ratepayers of Hulme in Manchester presented a petition to the Parks and Cemeteries Committee in 1913, asking for a park and pointing out that of the 1,315 acres of parks in Manchester, none were located in Hulme (MA, Parks and Cemeteries Committee minutes, Volume 32, p. 215). In 1914, Liverpool City Council moved a motion to instruct the Parks and Gardens Committee to provide a children's playground in the west part of North Scotland ward, one of the most congested in the city (LA, Parks and Gardens Committee minutes, Volume 28, p. 56). The committee replied that no suitable vacant land was available in that area. In a further communication on the subject, the city surveyor reported that the only method of providing suitable space would be to demolish some houses (LA, Parks and Gardens Committee minutes, Volume 28, p. 108). This was a familiar dilemma for many local authorities—poorer and more congested areas had no available land to provide for public recreation, and wealthier areas had more possibilities due to less dense housing provision.

The purchase and laying out of a public park in a particular area of a town or city was often believed to accrue certain advantages to that district and the surrounding area. These benefits exceeded the improvements to public health and the provision of recreational facilities. It was hoped that a public park would raise the value of land in an area and could potentially contribute to the formation of a new and desirable suburb. In some cases, the proximity of a large public park was used to develop or complement an existing municipal housing estate. In the 1904, Manchester City Council built its first such estate at Blackley, adjacent to Heaton Park. This estate was expanded considerably by the demands of the housing boom of the 1930s. Similarly, in Liverpool, Norris Green Park (16 acres) was opened in July 1933 next to the Norris Green estate of some 7,000 council homes, which had been developed during the 1920s. The intention was to provide recreational amenities for the residents of a 'brand new township' (O'Mahony, 1934, pp. 25–27). Lord Derby donated the land for the park on the condition that no houses were built on the site.

The Victorian parks' carefully planned, undulating walkways and the encouragement of gentle promenading and rational recreation were not entirely successful or consistent. Freely available public spaces attracted all who had the means to get there, and this often included people who flouted the regulations from the outset (Lambert, 2005). Changing definitions of 'the people' drove the emergence of such concepts as active citizenship and an increasing emphasis on individual social responsibility towards the end of the nineteenth century (Joyce, 1991, p. 191).

Parks and the Citizen: Public Leisure in the Twentieth Century

It has been suggested that citizenship in the Edwardian period was refocused away from the urban arena and onto the empire (Beaven and Griffiths, 2008). This study contests this view and demonstrates that active urban citizenship remained a potent social force in the landscape of the urban park. Such parks offered the opportunity to both establish and display not only a sense of civic pride in the city but also pride in the collective ownership of that space. Municipal parks, therefore, represented a place where urban citizenship could be continually forged and contested, both by park authorities and by park visitors.

The Edwardian recreational park, however, did facilitate different kinds of visitors—those who could concentrate on their own individual needs and interests and did not depend on being in company. This individual usage allowed visits to the park to be dictated by the free time schedules of individual visitors and permitted park-keepers to divide up each day into sections to offer facilities to different types of visitors—referred to as temporal zoning (Cranz, 1980). This was in an era when even working people had an increased amount of leisure time (Beaven, 2005). Indeed, it is possible to argue that recreation was beginning to be seen not as the opposite of work, as the Victorians believed, but as a civic duty, which complemented work and enhanced the individual's ability to function effectively in the workplace (Bailey, 1998).

A connection was made between civic pride and social citizenship in which the municipality assumed responsibility for the welfare of all citizens. The corollary of this is that city dwellers reciprocated in accepting the care of the urban environment as a part of their civic duty. Citizenship was an ambiguous term before the 1870s and encompassed potentially all of those who had a general interest in the welfare of the nation (Rodrick, 2004). From the late Victorian period, we find the model of citizenship becoming more proactive and socially aware. The needs of the empire were undoubtedly to become more significant as the twentieth century advanced—the use of Heaton Park as a training camp for the Manchester Regiment prior to their deployment in World War I demonstrates that the park evolved into a space that could accommodate such imperial needs while continuing to function as a public leisure space (Stedman, 2004). Thus, imperial, national and local citizenship could co-exist and were not mutually exclusive.

It has been argued that, by the end of the nineteenth century, public leisure facilities reinforced the desire for class exclusivity as a result of the appropriation of formerly aristocratic pursuits, such as hunting, by the middle classes, the invention of class-specific sports, like golf and tennis, and the imposition of a middle-class ethos on sports such as rowing and athletics (Cunningham, 1980). While this may be difficult to prove, it does provide an explanation for the increasingly class-bound nature of leisure at the end of the nineteenth century and would have militated against the ideal of recreation as a tool for unifying social classes advocated in the 1840s. It also marks the gradual refinement of ideas such as rational recreation, elements of which still persisted. Cardiff's Councillor Meyrick observed in 1902 that Roath Park was 'not simply for the man who wore broadcloth, not for the well-bonneted women; it was just as much for the man who lived a colourless life in the slums' (Pettigrew, 1932, Volume 6, p. 53).

There was an increasing tendency early in the twentieth century to conceive of poverty as a national problem characterised by the need for physical efficiency. There had been much concern about the poor physical condition of army recruits from the industrial cities during the Boer war. This had led to an acknowledgement that the people of Britain were an important national resource who needed to be nurtured and encouraged towards the peak of physical fitness (Searle, 1971). The provision of facilities for physical exercise in municipally owned parks was a consequence of this perception of the need to maintain levels of physical fitness among the population. Exercise facilities in public parks were not exclusively an Edwardian idea—the three original public parks in Manchester and Salford all had gymnasia (Salford's Peel Park had archery butts). The provision of this kind of equipment was an acknowledgement that parks were not simply open spaces for polite perambulations, but had a more pragmatic purpose. It has been argued that the Victorians tended to see leisure time as a contrast to idleness and as a valuable entity that should not be wasted (Rodrick, 2004). Rational recreation also included a purposeful dimension and an acknowledgement of the connections between healthy bodies and healthy minds.

The role of the urban park in the whole civic landscape was also now being considered. Liverpool City Council proposed the idea of creating a ribbon of parks around the city as early as 1850, but it lacked the necessary powers to raise the money to fund the scheme (Layton-Jones and Lee, 2008). The operation was partially realised between 1868 and 1872 with the creation of Newsham, Stanley and Sefton Parks to the north, east and south of the city (Layton-Jones and Lee, 2008). The Tory MP W. J. Bull proposed a similar idea for London. Bull suggested buying land adjoining London's parks to create a 'green girdle' that would run around the perimeter of the city. F. J. Holmes, writing in The *Quiver*, suggested extending this idea in other British cities, such as Newcastle and Sheffield, to create a 'glorious girdle of rurality' (January 1902, p. 192). Few of these suggestions were realised, but they are indicative of a new way of conceiving of the function of the park in the city as a whole and not merely as a contrast to it (notwithstanding Holmes's reference to 'rurality').

Attempts were also being made to connect the urban park to the history of its city in an explicit way. This contributed to the idea of the public park as a didactic space but also as one in which an appreciation of the city's history could be fostered. This was best exemplified in Manchester, where a proposal was made to move the Greek classical façade of the old Manchester Town Hall (built 1822–1825) to a city park in 1912. A campaign to support the saving of the colonnade was undertaken by some prominent individuals, such as the bishop of Salford, Louis Casartelli, the artist Sir Lawrence Alma-Tadema and Middleton architect Edgar Wood.

Despite the decline in interest in the Gothic Revival style at this time, feelings were still mixed about the historical significance and relevance of classical styles of architecture. A rediscovery of the English Baroque tradition of Wren and Vanbrugh during the Edwardian period did not result in a widespread return to classical style per se (Service, 1977). However, both the Victorians and Edwardians did retain a strong sense of affinity with the history and culture of ancient Greece, as Turner has demonstrated (Turner, 1981). He suggests that this can be explained in part by the Greeks' association with the birth of democracy, an ideal still cherished in the early years of the twentieth century. The links between the façade and the old town hall therefore had even more resonance for the civic authorities and help to explain their desire both to preserve and to display it.

In May 1912, the committee resolved to erect the colonnade at Heaton Park. Half of the estimated £2,000 cost of relocating the façade was to be met by the Corporation and the other half by public subscription. The symbolic nature of this decision cannot be underestimated. Stobart has argued that town halls have functioned as important symbols of municipal authority (Stobart, 2003). The use of a classical façade of a former town hall in this manner served as a reminder of the civic history of Manchester. It re-emphasised the public ownership of the park and the civic vision of those instrumental in the purchase. The façade was to act as a potent symbol of the history of the city and those who served it and created it. Relocating the façade to the park moved a part of Manchester's civic history into what had previously been a privately owned space developed by generations of one aristocratic family. The removal of the town hall façade to Heaton Park was an attempt both to preserve an element of Manchester's civic and architectural history and to connect the park visitors directly to their own history and that of their city. A newer form of park history was emerging that could co-exist with the park's original history, but that had a different meaning for its visitors and served to legitimate not only the municipal owners but all of the people of the city (Lowenthal, 1985).

A desire to preserve the past (even the relatively recent past represented by the façade) had begun to gather pace during the Victorian era and this continued into the Edwardian period. During the same time as the debate about the preservation of the old town hall colonnade, Lord Curzon purchased the fifteenth-century Tattershall Castle in Lincolnshire, which was to be restored and opened to the public (*Manchester Courier*, 1912). A sense of national pride in Britain's heritage was beginning to establish itself, accompanied by the idea that the past was worth preserving for more than mere aesthetic reasons. The National Trust had been established in 1893 and acquired its first property, four and a half acres of cliff land, in Wales in 1895 (Waterston, 1994). The original town hall building itself was not considered for preservation in its entirety, and the transfer of the façade to Heaton Park meant a loss of its original context. Nevertheless, it meant that the colonnade was preserved for the public and its consequent visibility gave any passer-by the opportunity to gain an immediate impression of the past (Lowenthal, 1985). Linking park visitors so strongly to a reminder of the city's past could be interpreted as an attempt to make people feel like custodians of their own civic history.

By opting to place the façade of the old town hall at Heaton Park, Manchester City Council was inviting park users to gaze on the object, to appreciate its heritage value and to use it to remember the city as it had been (Figure 2.4). It also formed a useful physical connection between the park and the city centre, four miles away, where the façade had been originally located. Public parks could, therefore, function as repositories of civic memory—reminders of what had once been and of a history that had once been in existence.

Other municipal authorities also considered similar exercises in civic memory in their public parks. Hull's Parks and Burials Committee suggested the erection of two stone arches at East Park in the city in 1921. The arches had originated from Newark Castle and had latterly been at the entrance to Holderness House in Hull, which was located near to East Park (HHC, Parks and Burials Committee, Volume 6, p. 49). The committee also brought the old Hull Town Hall staircase to East Park in the same year with a view to installing it but reconsidered their decision due to the cost of so doing (HHC, Parks and Burials Committee, Volume 6, p. 50).

The Edwardian period produced a new understanding of social democracy that emphasised the idea of the urban community and good citizenship (Meacham, 1994). This vision of democracy was defined by the

OLD TOWN HALL PORTICO, HEATON PARK, MANCHESTER.

Figure 2.4 Portico of the Old Town Hall at Heaton Park, Manchester.

Source: © Friends of Heaton Hall postcard collection.

harmony between nature and the individual and one that worked to idealise the past and improve on the present. The idea reached its artistic high point in the garden city movement of Raymond Unwin, Barry Parker and Ebenezer Howard during the early years of the twentieth century. The garden city was specifically designed to merge the country and the city and to encourage communal activities, such as tennis and bowling. Here, amenities were a right, not a privilege, and their proper use was a cornerstone of good citizenship (Meacham, 1994).

The early decades of the twentieth century were suffused with ideas of active citizenship, which included a commitment to good physical and moral health (Harris, 1993). Trade unions, co-operatives and friendly societies all provided opportunities and models of good citizenship, but this could also be extended to the role of public parks at this time. A healthy citizenry contributed to a healthy nation and formed the building block of a well-ordered society (Harris, 1993). It represented a shift in the ways in which parks were viewed and used in an urban environment, from attempts to impose ideas about rational recreation onto the working classes to an emphasis on community and a shared social responsibility for public amenities. The environment of a public park could offer the opportunity to develop not just physical health but also a sense of public spiritedness and civic identity. This can be seen in the use of public parks by the Boy Scout movement, one of whose primary aims was the development of citizenship skills (Warren, 1986). These activities were connected to emergent ideas about citizenship and collective responsibility for one's surroundings—a substantial move away from the Victorian idea of parks as patrolled by park-keepers and attendants who bore sole responsibility for the park's upkeep and maintenance.

In part, many of the decisions about how to develop public parks were a reflection of a broadening definition of public health, away from specific matters such as sanitation and slum removal and towards issues such as recreation and physical fitness. This manifested itself in organisations like the Manchester Physical Health Culture Society and Leeds's Everywoman's Health Movement devoted to promoting outdoor sports and physical development. Public parks offered a location where the city and the citizen could thus develop in tandem—to 'become a self-governing member of a self-governed community' (Dagger, 1981, p. 717). This emphasis on the community and the explicit link between health and well-being marks a transition from the Victorian middle-class moral imperialism of rational recreation to a more general concern with the health of the population as a whole.

The idea of the community was gradually expanded in the early decades of the twentieth century to include the recreational needs of women and children. By this time, many local authorities began to acquire smaller parcels of land, often in overcrowded areas to be developed as recreation grounds (or playgrounds). These spaces did not offer the facilities of the larger parks but were intended to ameliorate the drab streets and to provide basic recreational environments for deprived children. Leeds City Council began to develop such spaces from 1905, often taking over plots of land from charities and trusts (WYAS, Leeds Parks Committee minutes, Volume 1, p. 16). Chapel Allerton Park was a mere 6.5 acres and was located close to an estate of terraced houses. Similarly, Cardiff acquired such children's playgrounds from 1908 (Pettigrew, 1931, Volume 5, p. 40 passim). These spaces were conceived of as an alternative to children aimlessly hanging around on the streets and demonstrate the continuing belief that open space was preferable to urban streets and lanes.

Many of these recreation grounds were small in size and concreted and contained basic equipment, such as swings and seesaws. An official guide to parks in Leeds commented that their intention was to be used by children 'instead of being left to infest the street corners of the district' (Allsop, 1906, p. 215). The idea of the children of the urban poor as an infestation has connotations of disease and infection and serves as a reminder that the poor were perceived as having different and often quite separate recreational requirements. This was akin to the place accorded to women in parks.

The Victorians had prized the presence of women in public spaces such as parks as women were believed to be positive role models who encouraged good behaviour in others. Contemporary photographs suggest that municipal parks were popular with women in the early twentieth century, but due consideration of their specific recreational needs was not a priority at this time. Participation in such sporting activities as cycling was regarded as unladylike, leading women to be welcomed in public parks more for their stabilising influence than their ability to make active use of the facilities (Cranz, 1980). However, this situation did not persist in the longer term, mainly due to women's desire to actively participate in sports such as tennis and to the growing acceptance of at least some sports as permissible for women.

Women had been employed in parks since 1912 to instruct schoolchildren in safe play during the summer months, again reinforcing the perception of women as positive role models and emphasising their stabilising and didactic influence (*Manchester City News*, 1912). This also underlines the importance of teaching young children how to be good citizens through the use of appropriate role models, many of whom were teachers (*Manchester Guardian*, 1935). Teachers also had the advantage of knowing a wide range of games and being able to get the best out of the children.

In 1915, William Wallace Pettigrew, general superintendent of Manchester Parks, was invited by the Lancashire and Cheshire Committee for the Employment of Women to establish a training scheme in horticulture in public parks for the duration of the war (MA, Parks and Cemeteries Committee minutes, Volume 35, p. 224). Such training schemes were not unusual for young men, but this was the first time such opportunities had been aimed specifically at women. Six women were being trained at Heaton Park in September, and Pettigrew reported great interest in the project from the local newspapers, which had been asking for photographs of the trainees, underlining the novelty of the enterprise (MA, Parks and Cemeteries Committee minutes, Volume 35, p. 224). The specially designed syllabus, overseen by Pettigrew, had resulted in the women being 'thoroughly in earnest and taking quite an interest in the work' (MA, Parks and Cemeteries Committee minutes, Volume 35, p. 224).

The training scheme continued throughout the war years and resulted in many of the trainees successfully completing the course and gaining employment in parks around the country. Similar initiatives were replicated in many British public parks for the duration of the war (WYAS, Leeds Parks Committee minutes, Volume 2, p. 165). Some cities, however, such as Cardiff, did not employ any women in this capacity at all (Pettigrew, 1932, Volume 6, p. 6). The Manchester scheme ceased after the war, however, suggesting that such initiatives were regarded as a temporary wartime aberration and not a long-term commitment to the training of women in horticulture.

The leisure needs of other users were also being considered, and this demonstrates that parks were not just used as recreational spaces but offered other comforts, such as protection from the elements. Shelters built in parks had often used specific groups of people, particularly elderly men. The shelter in Manchester's Queen's Park was known as the Queen's Park Parliament due to the informal political nature of the discussions of those who gathered there regularly (MA, Parks and Cemeteries Committee minutes, Volume 44, p. 31). In 1933, the Parks and Cemeteries Committee was asked to provide a similar shelter for the Debdale Street Recreation Ground in Openshaw in the city, especially for use by those who 'have not many home comforts and have therefore to spend most of their time outdoors' (MA, Parks and Cemeteries Committee minutes, Volume 49, p. 55).

While the recreational needs of groups such as women were slow to be recognised in their own right, the impetus of the public parks movement changed from the Victorian rational recreation to Edwardian foregrounding of the active citizen. While this made public parks potentially more democratic and inclusive spaces, it also raised questions about the future developments of these parks in the context of the wider city. The emergence and popularity of private commercial forms of entertainment, such as music halls and cinemas, meant increasing pressure on public parks to compete as part of a general regimen of public health and leisure activities.

Healthy Cities, Healthy Citizens

The introduction of the 1925 Public Health Act enabled local authorities to rent out portions of public parks to local cricket and football clubs and to charge the public for admission to watch matches (Public Health Act, 1925, Section 69). Restrictions continued to be placed on the use of public parks for the purposes of entertainment. Costume concert parties (or 'Pierrot' parties) were not legalised by the Act, thus depriving many parks authorities of a reliable and popular form of income and preventing the staging of plays that required costumes and scenery.

The acquisition of large-scale land for public parks remained a goal of many municipal authorities during the early decades of the twentieth century—Manchester City Council received the donation of the 250-acre Wythenshawe Park from Ernest and Shena Simon in 1926, which inspired the development of a garden city suburb in that part of the city. But the emphasis also began to change in favour of the addition of smaller parcels of land, often referred to as recreation grounds (or 'pocket parks') and frequently located in inner city areas that had previously been overlooked as far as the provision of open space.

These were often acquired with the needs of working-class children in mind and imaginative steps were taken to take advantage of even the smallest of spaces. Cardiff Corporation acquired the tenancy of Tyndall Street playground in a congested area of the city in 1921. While some supervision of the playground was attempted by priests from an adjoining church, the chains were stolen from a giant stride (a telegraph pole with ropes attached that rotated and lifted one off the ground) that had been donated by a local councillor, and the tenancy was eventually terminated in 1926 (Pettigrew, 1931, Volume 5, p. 106).

Cardiff Corporation acquired the two acres of Waungron Common as a recreation ground in 1923. It was improved by the Parks Committee with the addition of hedges, fences and trees and reserved for young children only (Pettigrew, 1931, Volume 5, p. 44). The 1.75 acres of Ely recreation ground were acquired in 1926 and equipped with hard tennis courts and a bowling green. The opening ceremony was performed by the chair of the Parks Committee serving a tennis ball on one of the courts (Pettigrew, 1931, Volume 5, p. 86). Manchester City Council demolished St John's Church in the city centre in 1931 to provide a children's playground (O'Reilly, 2013). Stanley Park in Liverpool opened a children's garden in 1926, which contained statues inspired by popular children's stories. Sefton Park erected a statue of Peter Pan in 1929 and made available a series of life-size pirate ships on which children could play (Layton-Jones and Lee, 2008).

Many of the landscapes acquired by local authorities during the early decades of the twentieth century were not as carefully designed and planned as the early parks had been. Cardiff Corporation was given the 42 acres of Plymouth Wood by the Earl of Plymouth in 1923, which was kept as woodland with the addition of a few pathways (Pettigrew, 1931, Volume 5, pp. 48–58). These wilder landscapes were identified as more 'natural' than planned and designed parks. The *Western Mail*, in its account of the opening of Plymouth Wood, stressed the 'natural beauties' of the wood and that the new park's visitors were now 'part owners with other citizens' (Pettigrew, 1931, Volume 5, p. 53). Such landscapes were also less expensive to establish and maintain in the longer term, as their wild nature was believed to be both an advantage and an attraction for visitors.

This also emphasises the break with the Victorian park, which was intended to be a contrast to the surrounding cityscape. Now, citizens were expected to be able to make their own use of the park according to their interests and their circumstances. The twentieth-century park sought to integrate its landscape with the urban environment and to carve out its own future, one element of a more unified approach to public health. Active citizenship and social responsibility had united the municipal public parks with the wider urban environment and integrated them more fully into the city.

Conclusion

The Edwardian park offered both continuity and a breach with its Victorian forebear. Clearly, there was not an abrupt transition from the Victorian attitude to the Edwardian approach but rather a gradual repositioning of thought. Much of this occurred at the level of the municipal authorities. The city was increasingly conceived of as a social body whose future prosperity depended on the health of each component part. New powers to legislate at a local level gave the municipality a degree of autonomy over its own affairs but also encouraged a more proactive approach to city management.

With increasing amounts of free time available for leisure, citizens took full advantage of the municipal park, but those who benefited most had both the recreational skills and the access to the spaces. The working classes remained on the periphery and the needs of particular groups of users, such as women, were yet to be fully met or even recognised. Parks continued to function as social arenas where models of good behaviour and citizenship could be observed and imitated. The restrictive atmosphere of the Victorian park gradually eased as responsibility for moral and physical rectitude passed from the park-keeper to the individual visitor. The effect of this was a transfer of emphasis from the passive strollers (whose needs were still accommodated) to the active users whose various recreational needs could be served simultaneously. A new type of diverse cityscape was now capable of serving a new kind of citizen—one whose demands for public leisure facilities were only beginning.

There has been an overemphasis on the Victorian park at the expense of later Edwardian advances and too much emphasis on rational recreation and social control, which offer a limited view of the practical usage of parks. The concept of rational recreation does not allow for unintended uses made of these parks for meetings and games and offers no prospect of the visitor's individual enjoyment of the space. While Edwardian public parks were an evolution of those that originated in the Victorian period, they also developed their own character and established new ways for some people to spend their increasing amounts of leisure time.

New types of public space, such as parks, allowed new types of normative behaviour to emerge, often that which challenged the prevailing social norms. This is the subject of the next chapter.

Acknowledgements

Parts of this chapter are derived from articles published in *Urban History* (February 2013) © Cambridge University Press (available online: https://doi.org/10.1017/S0963926812000673) and *Landscape History* (October 2017) © the Society for Landscape Studies (available online: https://tandfonline.com/doi/abs/10.1080/01433768.2017.1394066).

References

Allsop, A. J. (1906). *The Official Handbook to the Public Parks of Leeds and Kirkstall Abbey.* Leeds: John Waddington.

Bailey, P. (1998). *Popular Culture and Performance in the Victorian City.* Cambridge: Cambridge University Press.

Beaven, B. (2005). *Leisure, Citizenship and Working Class Men in Britain 1850–1945.* Manchester: Manchester University Press.

Beaven, B. and Griffiths, J. (2008). Creating the Exemplary Citizen: The Changing Notion of Citizenship in Britain 1870–1939. *Contemporary British History,* 22(2), 203–225.

Briggs, A. (1963). *Victorian Cities.* London: Penguin.

Burt, S. (2000). *An Illustrated History of Roundhay Park.* Leeds: S. Burt.

Concilio et Labore (1912, 6 May). *Manchester Courier,* p. 6.

Conway, H. (1991). *People's Parks: The Design and Development of Victorian Parks in Britain.* Cambridge: Cambridge University Press.

Cranz, G. (1980). Women in Urban Parks. *Signs: A Journal of Women in Culture and Society,* 5(3), 79–95.

Cunningham, H. (1980). *Leisure in the Industrial Revolution.* London: Croom Helm.

Curl, J. (1975). The Architecture and Planning of the Nineteenth Century Cemetery. *Garden History,* 3(3), 13–41.

Curl, J. (1983). John Claudius Loudon and the Garden Cemetery Movement. *Garden History,* 11(2), 133–156.

Dagger, R. (1981). Metropolis, Memory and Citizenship. *American Journal of Political Science,* 25(4), 715–737.

Erikson, K. T. (1960). Notes on the Sociology of Deviance. Paper presented at the American Sociological Association Conference, New York.

From Graveyard to Playground (1935, 28 May). *Manchester Guardian,* p. 8.

Games With a Difference (1935, 30 August). *Manchester Guardian,* p. 10.

Garrard, J. (1983). *Leadership and Power in Victorian Industrial Towns 1830–1880.* Manchester: Manchester University Press.

Harris, J. (1993). *Private Lives, Public Spirit: A Social History of Britain.* Oxford: Oxford University Press.

Holmes, F. J. (1902, January). Green Girdles Around Great Cities. *The Quiver,* p. 192.

Hull History Centre, Kingston-Upon-Hull, Minutes of the Parks and Burials Committee, TCM/2/14-44./6.

In the Park (1876, 28 July). *Hull Packet and East Riding Times,* p. 8.

In the Parks (1902, 16 May). *Manchester Courier,* p. 10.

Jacobs, J. (1964). *The Death and Life of Great American Cities.* London: Penguin.

Joyce, P. (1991). *Visions of the People: Industrial England and the Question of Class 1840–1914.* Cambridge: Cambridge University Press.

Lambert, D. (2005). *The Park Keeper.* Swindon: English Heritage.

Laski, H. (1936). The Committee System in Local Government. In H. Laski, W. I. Jennings and W. A. Robson (Eds.). *A Century of Municipal Progress: The Last Hundred Years* (pp. 82–108). London: George Allen and Unwin.

Layton-Jones, K. and Lee, R. (2008). *Places of Health and Amusement: Liverpool's Historic Parks and Gardens.* Swindon: English Heritage.

Liverpool Archives, Liverpool, Parks and Gardens Committee Minute Books, 352/MIN/PAR/1/.

Lowenthal, D. (1985). *The Past Is a Foreign Country*. Cambridge: Cambridge University Press.

Manchester Archives and Local Studies, Manchester, Parks and Cemeteries Committee Minute Books, GB127.Council Minutes/Parks and Cemeteries/1-53.

Maver, I. (1998). Glasgow's Public Parks and the Community, 1850–1914: A Case Study in Scottish Civic Interventionism. *Urban History*, 25(3), 323–347.

Meacham, S. (1994). Raymond Unwin 1863–1940: Designing for Democracy in Edwardian England. In S. Pedersen and P. Mandler (Eds.). *After the Victorians: Private Conscience and Public Duty in Modern Britain* (pp. 79–104). London and New York: Routledge.

Municipal Affairs and Press Criticism (1897, 24 April). *Manchester City News*, p. 4.

O'Mahony, M. (1934). *Official Handbook: The Parks, Gardens and Recreation Grounds of the City of Liverpool*. Liverpool: Liverpool City Council.

O'Reilly, C. (2009). *Aristocratic Fortunes and Civic Aspiration: Issues in the Passage of Aristocratic Land to Municipal Ownership in Later Nineteenth and Early Twentieth Century Manchester With Particular Reference to Heaton Park* (Unpublished PhD Thesis), Manchester Metropolitan University, Manchester.

O'Reilly, C. (2013). 'We Have Gone Recreation Mad': The Consumption of Leisure and Popular Entertainment in Municipal Public Parks in Early Twentieth Century Britain. *International Journal of Regional and Local History*, 8(2), 112–128.

Organised Play for Children (1912, 8 June). *Manchester City News*, p. 6.

Penny, N. (1974). The Commercial Garden Necropolis of the Early Nineteenth Century and Its Critics. *Garden History*, 2(3), 61–76.

Pettigrew, A. A. (1926–1932). *The Public Parks and Recreation Grounds of Cardiff*. 6 Volumes. Unpublished.

Pettigrew, W. W. (1937). *Municipal Parks: Layout, Management and Administration*. London: The Journal of Park Administration.

Public Health Act, 1848, 38 & 39 Vict., c. 63.

Public Health Act, 1875, 38 & 39 Vict., c. 55.

Public Health Act, 1925, 15 & 16, Geo V, c. 71.

Report from the Select Committee on Public Walks With the Minutes of the Evidence Taken Before Them, 1833. Parliament: House of Commons.

Rodrick, A. (2004). *Self-Help and Civic Culture: Citizenship in Victorian Birmingham*. Aldershot: Ashgate.

Searle, G. R. (1971). *The Quest for National Efficiency: A Study in British Politics and Political Thought 1899–1914*. London: Blackwell.

Service, A. (1977). *Edwardian Architecture: A Handbook to Building Design in Britain 1890–1914*. London: Thames and Hudson.

Seton, F. E. (1928, 27 October). Garden Design. *The Spectator*, p. 572.

Stedman, M. (2004). *Manchester Pals*. Barnsley: Pen and Sword Books.

Stobart, J. (2003). Identity, Competition and Place Promotion in the Five Towns. *Urban History*, 30(2), 164–182.

Sullivan, J. J. (1915). *Illustrated Handbook of Manchester City Parks and Recreation Grounds*. Manchester: Manchester Parks and Cemeteries Committee.

Turner, F. (1981). *The Greek Heritage in Victorian Britain*. London and New Haven: Yale University Press.

Twist, C. (2000). *A History of the Liverpool Parks*. Southport: Hobby Publications.

Walton, J. (1987). *Lancashire: A Social History 1558–1939*. Manchester: Manchester University Press.

Warren, A. (1986). Sir Robert Baden-Powell, the Scout Movement and Citizen Training in Great Britain 1900–1920. *The English Historical Review*, 101(399), 376–398.

Waterston, M. (1994). *The National Trust: The First Hundred Years*. London: The National Trust.

West Yorkshire Archives Service, Leeds. Parks Committee Minute Books. LLC47/1/.

The Working Classes and Public Parks (1869, 7 December). *Leeds Mercury*, p. 7.

Wyborn, T. (1995). Parks for the People: The Development of Public Parks in Manchester. *Manchester Region History Review*, 9, 3–14.

3 Parks as Social and Political Landscapes

Introduction

Urban parks truly became social spaces only through their regular usage. However, how they were to be used and by whom remained contentious. Parks were frequently the scene of political marches and demonstrations (by the labour movement, suffragettes, temperance adherents), peopled by vagrants, unruly children, courting couples and sexual predators. Patrolling and regulating these parks were a considerable challenge and one that the park employees, such as park-keepers, often struggled to meet. Only through an examination of the competing and contested uses to which these parks were subjected by all of their users can we begin to understand the passions they provoked. Parks were a showcase for both best clothes and best behaviour, and the Victorian idea of these spaces as a form of social control persisted albeit in subtler forms until well into the twentieth century. This chapter focuses on the contested nature of such public spaces and is informed by the views of those who used them and by local commentators and publications, as well as by those who ran and worked in them.

Parks, Citizenship and Normative Behaviour

Britain's urban parks were sites of social and political conflict. Initially conceived of as places where different social classes could mingle and where the lessons of urban citizenship could be learned, they developed a reputation as socially segregated and controversial spaces. Many of the strict social divisions of the Victorian age were imported directly into the park landscape. Instead, therefore, they often inadvertently reproduced the same inequalities they were established to abolish.

Mark Philips (1800–1873), the Manchester MP and local parks supporter, tried to capture the spirit of egalitarianism with respect to access to open space, when he wrote that 'I do not know why the weaver and the mechanic should not cultivate a taste for flowers as the first nobleman in the land' (Lasdun, 1991, p. 163). The sense of shared ownership represented by the phrase 'people's parks' was emphasised at the 1846 opening of Philips Park in Manchester, partly funded by the Philips family: 'I urge upon every member of your families and upon every friend, coming into these pleasant places, that they have an individual property in every tree, plant, shrub and walk' (*The Manchester Times*, 1846). The latter comment is especially noteworthy in terms of the emphasis on collective ownership of the space. This was to have profound and sometimes difficult results, which will be discussed later in the chapter.

The genesis of the term 'people's park' (which was later transformed into the rhetorical flourish of 'parks for the people') is usually attributed to Frederick Law Olmsted, who used the phrase on a visit to Birkenhead Park on the Wirral in 1851. He observed that 'the privileges of the garden were enjoyed about equally by all classes' and referred to it as a 'People's garden' and 'the People's own' (Beveridge and Hoffman, 1997, p. 71). This rhetoric was subsequently adopted in many British cities and used by civic dignitaries and municipal representatives on the occasion of the ceremonial opening of public parks. Speaking at the opening of Sefton Park, Liverpool in 1862, the park designer Charles Melly proclaimed that 'It is . . . to the hard-working honest men living in this neighbourhood that I consign the care of this playground' (*The Liverpool Mercury*, 1862). The Lord Mayor of Bradford echoed this sentiment in 1878 at the opening of Horton Park: 'As the rich man had his park . . . so the poorest man, when he entered these gates could say: "This is my park"' (*The Leeds Times*, 1878). The phrase 'people's park' also began to appear in reader's letters on the subject of the provision of municipal parks. One letter writer, signed Alpha, wrote of 'almost a craze in Belfast on the subject of parks for the people' (*The Belfast Newsletter*, 1892). Cardiff's councillor Meyrick observed in 1902 that Roath Park was 'not simply for the man who wore broadcloth, not for the well-bonneted women; it was just as much for the man who lived a colourless life in the slums' (Pettigrew, 1932, p. 53).

The implication of a 'people's park' was that they were for all of the people, for their interest and enjoyment and the improvement in their health. The concept of 'the people' was originally intended to be classless and inclusive of all social classes. Many of Britain's largest municipal parks were in predominantly middle-class areas in cities such as Manchester, Leeds and Liverpool. Thus, the phrase 'parks for the people' bore little reality to the lives of many of the urban poor who were no better off than they had previously been.

'Parks for the people' was rarely used during municipal debates on the acquisition of land for public parks and was found more commonly in the pages of local newspapers as part of the case for civic improvement. Its all-inclusive allusion emphasises the Victorian desire to unite all social classes in the modern city as a solution to the problems of social order and urban overcrowding. However, the idea of 'the people' is problematic in its assumption of a shared destiny and interests. In practice, 'the people' do not constitute a homogenous group and cannot all occupy a space equally (Rodrick, 2004). Furthermore, Joyce has pointed out that shifting definitions of 'the people' resulted in the emergence of such concepts as the common good and social citizenship (Joyce, 1991). These terms embody a more active connotation than 'the people' and imply an urban dweller that is, potentially at least, actively engaging with his or her environment.

This sense of shared ownership and responsibility was variously felt by parks visitors. In fact, many people apparently misinterpreted this as an invitation to pick the flowers and to take home shrubs, bushes and bulbs for themselves. The notion of a 'people's park', therefore, exerted a powerful influence on the experience of that park and created a sense of entitlement in some cases. It is possible that these occasions of theft were a reaction to the perception of exclusion from the paid-for amenities at the parks and stemmed from a determination to recoup something from a visit or to provide a memory of a rare day out or were simply an opportunistic impulse. Griffin has identified such acts in rural locations as acts of protest or resistance, but we have little evidence about the reasons behind this behaviour in public parks (2008). Petty vandalism was a regular feature of both the larger parks and the smaller recreation grounds. Much of this may be attributed to boredom experienced by young boys. In 1882, three boys were observed in a Hull park putting a football into mud and then throwing it at a statue of Queen Victoria (HHC, Parks and Recreation Grounds Committee, Volume 6, p. 8). However, discretion was exercised in the response to these events. Four boys were accused of breaking the drinking fountain in Salford's Ordsall Park in 1899. As a result, they were given a fine by the local magistrate, which their parents were unable to pay. The parks superintendent recorded that 'all these people were very poor and very sorry that the damage had been done' and asked the Parks Committee to dismiss the case, which was done (SA, Parks Superintendent Report Books, Volume 4, pp. 102–103).

Much of the existing work on the history of the public park has emphasised its role in social control. Social control has been defined as 'the analysis of the processes that tend to counteract deviant tendencies' or attempts to manipulate or alter behaviour (Taylor, 1999, p. 421). This has most often been allied with the concept of 'rational recreation', an attempt to impose middle-class values onto working-class leisure habits. Parks have been approached as an important tool in the development of mechanisms to discourage the working classes from drinking and gambling and to reorientate their behaviour to more acceptable norms. Public parks were supposed to provide an environment in which this self-improvement could take place under the guidance of the middle class and according to their standards. Thus, the values of good citizenship could be transmitted and learned while enjoying genteel recreational pursuits.

Frequently, writers have stressed the primacy of 'rational recreation' in the context of the Victorian outlook, dominated by respect, sexual inhibition and a desire for conformity. In actual fact, this attempt to exert social control over the working classes continued well beyond Victoria and public parks remained a significant blueprint for citizenship and social learning well into the twentieth century. Care must be taken not to overemphasise the control element here—while there were frequent and ongoing attempts to regulate (and thereby exert control over) behaviour in public parks, much of this was a failure as users either ignored or broke the rules with impunity. At least some of this behaviour was not conscious or deliberate but opportunistic in nature, and speculative imputations are unavoidable in many cases due to lack of evidence. However, it is vital not to mistake 'archival silence' for passivity in respect of the various uses, intended and unintended, made by those who visited public parks (Stedman Jones, 1983, p. 78).

Much of the desire to control working-class leisure time resulted from its public nature—in either the street or the public house. However, it is important to recognise that many civic leaders made a careful distinction between 'rough' and 'respectable' working class (Stevenson, 1984, p. 341). In reality, therefore, it was the subset of the 'rough' working classes which was the focus of many attempts at social control. Etheridge has remarked on the use of music as a form of rational recreation, with brass bands 'at the forefront of the desire to give the working class an improving pastime' (Etheridge, 2014, p. 166). The fact that so much of the music played in public parks during their early years was performed by brass bands means that the bands and their environment made an ideal symmetry. Similar distinctions between 'rough' and 'respectable' were made in other didactic municipal environments, such as libraries, art galleries and museums (Hill, 2001). Intraclass social control was also practised—many of those who worked in public parks, especially in the unskilled occupations of patrolling and watching, were effectively policing members of their own class. This is discussed more fully in Chapter 4.

Attempts to regulate working-class behaviour were not new—factories and mills established strict rules governing their employees and it therefore seemed natural that middle-class employers should seek to regulate their workers' use of leisure time also (Thompson, 1963). Such attempts to impose top-down regulation were often resisted, and the history of our public parks represents, in part, a continuous struggle over what the space meant and how it should be used. Indeed, it is no coincidence that many of the towns and cities under consideration here were newly self-governing as a result of the 1835 Municipal Corporations Act. This initiative meant that new forms of normative behaviour were emerging in these newly autonomous places and questions were beginning to be asked about who controlled behaviour and what agendas were being pursued.

The fact that urban parks emerged at the same time as many of these new urban authorities were developing a police force is an indication that much thought was being given to how urban behaviour was displayed, learned and regulated. Such normative considerations provide a useful starting point for an examination of policies in respect of the regulation and control of parks at this time. Decisions about how to enforce these rules and how and if they should be altered over time are a sign of how dynamic and organic these spaces proved to be.

While there were many continuities between Victorian and early twentieth-century Britain in respect of the regulation of leisure and some unease about its prevalence among the working class, there was undoubtedly a gradual relaxation of attitudes towards working-class leisure towards the end of the Victorian period. Recreation became less a social evil than a civic necessity— something that refreshed workers and enabled them to perform better in the workplace (Bailey, 1998). This placed additional pressures on individuals, however, to take greater responsibility for their actions and behaviour while in public. This pressure continued to place at least some emphasis on the need to educate the workers and to provide them with suitable role models in public.

Public parks were also the location of a growing set of ideas about women's leisure and women continued to visit parks in considerable numbers. Why were parks so attractive to women? Clearly, many brought their children, which can be seen from contemporary photos, which show many women with groups of children. Women with children went with other women with children, often without men, as can be seen in this image outside Moor Park in Preston (Figure 3.1). It represented a rare opportunity to escape the domestic environment into the fresh air and to enable the children to run and play in a safe place. Some of the sporting facilities on offer were especially attractive to women, such as tennis and cycling. Edwardian fashions permitted women to ride bicycles in a manner that was socially acceptable and it rapidly became a popular pastime. Bicycles are in evidence as a mode both of transportation to the park and of getting around within it. By the Edwardian period, cycling was increasingly popular, both as a hobby and as a means of physical exercise among both men and women. Production and design innovations had made bicycles cheaper to purchase, lighter and more comfortable to ride, and cycling rapidly became an activity that transcended class, gender and age (McCrone, 1988).

Rarely has the issue of gender been considered in both the design and layout of the parks or in regarding those who actually used the parks and their facilities. The early twentieth century is a particularly fruitful time in which to examine the use of parks on a gendered basis due to the changing relationships between men and women and the growing acceptance of women as urban citizens in their own right. Birchall points out that 'in late nineteenth- and early twentieth-century cities, public spaces

Figure 3.1 Women at Moor Park, Preston, 1924.

Source: © The Francis Frith Collection.

allowed young working-class women to utilise a certain freedom that was available to them away from the constraints of family supervision existing closer to home' (2006, p. 243). While the perception of women as equal to men remained someway off, there was an increasing emphasis on directly addressing the leisure needs of women and children, especially those from the poorer classes. This did not always involve conceiving of women qua women but as *individuals* in their own right. The feminisation of leisure in the twentieth century illuminates a broader social trend of the acceptance of women as civically engaged contributors to society and consumers of public leisure—a parallel development to the rise of retail consumerism and the department store.

Women's opportunities for leisure pursuits had always been constrained by two elements—time and money (Langhamer, 2005). Women's leisure needs had not been recognised in their own right or had been subordinated to those of their husbands or their children. By the early decades of the twentieth century, women had begun to articulate their leisure needs more forcefully, albeit their participation was still perceived in quite limited terms. Manchester's parks superintendent William Wallace Pettigrew worried about the impact of women selling their own refreshments while bowling and playing tennis in the parks in 1927. He observed that 'with the advent of ladies as participants in park games, the afternoon cup of tea and its adjuncts of cakes and sweets soon made their appearance as they added to the sociability of the games' (MA, Parks and Cemeteries Committee minutes, Volume 44, p. 114). The explicit yoking together of female participation and sociability continued a tradition of viewing women's leisure through the narrow lens of traditional domesticated femininity, devoid of references to health or competitiveness.

There were also other more economic deterrents to women's ability to take full advantage of parks' facilities. In 1930, Sarah Laski, a newly elected Liberal city councillor, wrote to Manchester Parks and Cemeteries Committee to protest the high charges for refreshments in the city's parks. She wrote that:

> the mothers who have to prepare the refreshments for their families prior to a day's outing in Heaton Park are tired out before they start and it is this type of overworked mother who could have a real holiday if the charges were moderate and reasonable and they could take their little families into the refreshment rooms and enjoy what is provided there and be waited upon.
>
> (MA, Parks and Cemeteries Committee minutes,
> Volume 46, p. 134)

This plea for a particular class of mother reveals much about the failure of many larger parks to cater for their needs or to provide affordable opportunities for respite from the demands of family life. Laski's feelings were echoed by her fellow councillor Hannah Mitchell, who wrote that 'tired mothers who had trundled prams and "go-carts" all afternoon should not have to drag them home again before they could have a cup of tea' (1968, p. 207), reflecting her feeling that refreshment room facilities in public parks were too expensive and did not provide sufficient comforts for those with young children. The struggle of young mothers was clearly a scenario with which Laski was familiar as she was active in the Council for the Unmarried Mother and the Manchester and Salford Council for Social Service (Hunt, 2012). As a result of her letter, the Parks and Cemeteries Committee agreed to experiment with lower charges in the refreshment rooms.

Thus, public parks continued to function as sites of display of middle-class values and norms and emphasised the newer cultures of active citizenship and social responsibility (Steinbach, 2012). In some ways, this marked a return to the idea of a 'people's park' but with a growing importance attached to the exercise of communal responsibility for its upkeep. Authority was no longer solely vested in the figure of the park-keeper. From the late Victorian period, we find the model of citizenship becoming more proactive and socially aware. The needs of the empire were becoming more significant as the twentieth century advanced—the use of public parks for military training purposes during World War I, for instance, ensured that local, national and imperial citizenship needs could all be accommodated and were not mutually exclusive.

Parks and Political Meetings

From their earliest inception, municipal parks were used for the staging of political rallies and demonstrations. These events had been held in public squares and in meeting halls prior to the advent of public parks. Thus, any publicly accessible, open space became simultaneously an 'open-minded space' in which various social and political movements could agitate, express dissent and explore ideas (McKay, 2011, p. 15). Clearly, this aspect of their use ran counter to ideas about social control and parks as moral enclaves in which good behaviour could be learned. However, the nature of the meetings held in public squares was quite different to that in the parks. Parks were generally sites for organised demonstrations by recognised political groups, such as the Independent Labour Party (ILP) and the Women's Social and Political Union (WSPU). In contrast, public squares and other open spaces in the city were more often the site for spontaneous and less formal gatherings, such as those by groups of the unemployed.

The use of public parks for this purpose remained one of the more controversial aspects of public leisure during this period. Universal access to a public park could not be assumed in all cases. Many parks retained private areas that could not be accessed by the general public or sold off some land for the building of private houses or villas in order to defray the cost of the park's purchase. Some of the earliest parks, such as Liverpool's Princes Park (1842), had developed not as a public park but as semi-privatised commercial entity, with some areas open to the public and others not (Layton-Jones and Lee, 2008). Leeds's Roundhay Park was partly financed by the sale of land for building houses. This proved difficult as the site was located at some distance from the city centre (Burt, 2000).

Most civic authorities set aside a certain space in some of their parks (usually the larger ones) to accommodate political meetings. Manchester's leading Tory newspaper, the *Manchester Courier*, commenting on the lack of large open spaces in the city, opined that bigger public parks, like Hyde Park (a large park in Central London where political meetings were regularly held), could offer a place 'where a melancholy Jacques can rail at his fellow-men without interfering with their comfort' (*The Manchester Courier*, 1896). Parks were thus conceived of as places where political pressure could be relieved and controversial views expressed peacefully and under certain controlled conditions. Permission to hold political meetings had to be applied for in advance and were confirmed only if the local police force had no objection. The primary concern was to prevent these meetings interfering with the enjoyments of other parks users. Any failures to adhere to the rules pertaining to the holding of political meetings were taken seriously.

In Manchester, two local ILP activists, Leonard Hall and Fred Brocklehurst, were jailed for a month for holding a public meeting without permission in Boggart Hole Clough Park in 1896. The jailing of the two men had elicited widespread condemnation from organisations including the Manchester and Salford Trades and Labour Council and the South Manchester Liberal Association (MA, Parks and Cemeteries Committee minutes, Volume 16, p. 126). On 22 May 1896, Manchester City Council passed a bye-law banning the use of public parks for political meetings, raising the issue of what kinds of activities would be permitted in public spaces and how (and by whom) they should be policed. The 1896 bye-law was amended in January 1897 after the intervention of the home secretary, Sir Matthew Ridley, and political meetings were permitted in the city's parks (Ruff, 2000). Although the new bye-laws allowed such meetings subject to certain conditions, such as not raising money, tensions continued to prevail about the political content of such meetings. The ILP and the suffrage movement were now tolerated, and Manchester's largest parks saw audiences attend meetings organised by the WSPU in the summer of 1908.

Some parks authorities were more cautious about the use of public parks for organised political meetings. Liverpool's Parks and Garden Committee was adamant that the parks under its control should not be used for this purpose. Bye-law 30 of the parks regulations specifically prohibited the holding of any political meetings at all: 'No person shall sing, preach, lecture or take part in any religious service, public discussion or meeting for political or religious purposes in the parks' (LA, Parks and Gardens Committee minutes, Volume 20, p. 391). There were good reasons for this—Liverpool suffered from regular outbreaks of sectarian violence on its streets, especially between Protestants and Catholics. Anti-Catholic meetings were held in public places, such as outside St George's Hall in the city and near a disused mere called St Domingo pit in the largely Protestant area of Everton (Neal, 1988). Some violence had resulted from these meetings and the police had to intervene.

In December 1902, a proposal appeared in front of the Parks and Gardens Committee to allow political meetings in certain parts of Sefton, Newsham and Stanley parks in the city. A report to the committee from the head constable, Leonard Dunning, outlined the difficulties caused by these meetings and their inflammatory speaker, George Wise, and the advisability of allowing him to hold his meetings in public parks instead of in public spaces and waste ground around Liverpool. The committee voted in favour of allowing meetings to continue at St Domingo pit (which was Corporation property) but voted against allowing meetings in the three large public parks (LA, Parks and Gardens Committee minutes, Volume 23, p. 43).

The resistance to political meetings in Liverpool parks continued into the twentieth century and is a reminder of the political tensions that dominated the city. In April 1907, the south-east Liverpool branch of the ILP wrote to the Parks and Gardens Committee to protest the lack of meeting spaces in the public parks and to accuse the committee of failing to provide places for free speech for all citizens (LA, Parks and Gardens Committee minutes, Volume 24, p. 486). Another vote on the subject in January 1909 was lost seven to two, as was a proposal to spend one month examining the situation in other cities, such as Manchester, London and Birmingham (LA, Parks and Gardens Committee minutes, Volume 25, p. 345).

While in most cities and towns, political meetings were permitted according to certain preordained regulations, some local authorities continued to discriminate against the holding of political meetings or demonstrations by some groups. This was true of the co-operative movement in Manchester and the Socialist Labour Party in Leeds. The co-operative movement was not permitted to hold any meetings in Manchester's parks, while the Socialist Labour Party was not allowed to sell literature or to collect donations in parks in Leeds (WYAS, Leeds Parks Committee minutes, Volume 1, pp. 259–260). The holding of a meeting by the WSPU was banned in Preston's Moor Park in 1909 as it contravened the bye-laws (LRO, Health and Recreation Committee minutes, Volume 8, p. 26).

In 1901, a deputation from the Co-operative Society met with Manchester's Parks and Cemeteries Committee to discuss its situation and to press its case that it should be treated equally with other organisations, such as the Women's Social and Political Union, which held regular meetings in public parks in the city (MA, Parks and Cemeteries Committee minutes, Volume 28, p. 197). While the Co-operative Society was eventually permitted to hold its meetings from October 1909, its members were not allowed to make speeches advertising co-operation (ibid., p. 215). This was reaffirmed by a further vote in April 1910 (MA, Parks and Cemeteries Committee minutes, Volume 29, p. 123).

In June 1910, the Co-operative Union wrote to the Parks and Cemeteries Committee to protest this restriction, which was defended by the Manchester Retailer Trader's Association. Given the likelihood that many of the members of the city council were also members of this latter group, we can begin to see the origin of the discrimination against the Co-operative Union (Gurney, 1997). The controversy continued until June 1912, when the Manchester and Salford Equitable Co-operative Society asked for permission to use a lurry (lorry/ truck) displaying advertisements for co-operative goods at a gala in Alexandra Park (MA, Parks and Cemeteries Committee minutes, Volume 31, p. 208). The request was refused.

The granting of requests to hold Co-operative Society meetings with speeches was not uncontroversial in other British cities. The Leeds and District Grocers and Provision Dealers Association wrote to Leeds Parks Committee in 1912 to protest the use of its public parks 'for propagating co-operative principles' (WYAS, Leeds Parks Committee minutes, Volume 2, p. 34). Often, potentially controversial political meetings were rather tame in outcome. In Salford in 1925, a meeting of the Young Communist party was allowed in Peel Park but was 'sparsely attended', according to the parks superintendent (Salford Archives [hereafter SA], Parks Superintendent Report Book, Volume 8, p. 219).

Many of the meetings held in public parks during the early years of the twentieth century were organised by suffrage groups. It is notable that, from the 1880s onwards, more women were becoming politically active and they often had recourse to public parks to publicise their causes (Pugh, 1994). Emmeline Pankhurst founded the Women's Social and Political Union (WSPU) in her Manchester house in 1903. In 1908, the WSPU successfully applied for permission to the Parks and Cemeteries Committee to hold a series of public demonstrations at Heaton Park and other Manchester parks in favour of votes for women. Two demonstrations were held at Heaton Park—on 11 and 19 July, 1908. The demonstration on 11 July was echoed at Alexandra Park on the same date and was attended by almost 10,000 people. The larger demonstration on 19 July attracted an estimated 50,000 people (*Manchester Evening Chronicle*, 1908).

The WSPU speakers were accommodated on 13 platforms and included Mrs Pankhurst and her daughters Adela and Christabel and Mary Gawthorpe, a Leeds school teacher. While there was some heckling from 'rowdy youths and young men' the general organisation of the demonstration was praised by local newspapers and attendees (*Manchester Evening News*, 1908). These demonstrations had not always been peaceful—a suffragette rally in July 1906 held at Boggart Hole Clough park, in the same place where Hall and Brocklehurst had been arrested a decade earlier, broke up in disarray after 15 minutes when hecklers threatened the safety of the speakers in a manner described by the parks superintendent as 'very disorderly' (MA, Parks and Cemeteries Committee minutes, Volume 26, p. 43).

Women's use of the urban park as a political space and as an important component in the campaign to secure votes for women continued to evolve as in the early hours of 11 November 1913 a small bomb was detonated outside the cactus house in Alexandra Park, Manchester. The bomb caused £200 worth of damage, mostly to the glass of the building. The attack was immediately blamed on the suffragettes by the police, who allegedly found women's footprints near the building and bicycle tracks leading to the cactus house (*Manchester Guardian*, 1913). Similar incidents had taken place at Kew Gardens and Regent's Park in London in March 1913 when an orchid house and a refreshment room were burned (Rosen, 1974). The Manchester incident did not bear the hallmarks of these previous occasions when suffrage literature was left at the sites. Nevertheless, Manchester newspaper reports of the incident presented it as the work of the suffragists despite the circumstantial nature of the evidence. They all mentioned the presence of women's footprints but only the *Manchester Guardian* referred to the use of bicycles. Bicycles as a form of transport and recreation had become increasingly popular with women during this time, but the swiftness of the local press to attribute this event to suffragette action indicates the difficult nature of women's political protest and the narrow confines it was perceived to inhabit (McCrone, 1988).

The subject of temperance also resulted in several public demonstrations in Manchester parks in the early twentieth century. Manchester was an important centre of the temperance movement and the birthplace of the United Kingdom Alliance, whose annual general meetings in the city were well attended (Dingle, 1980). An estimated 250,000 people attended a temperance demonstration in June 1908, accompanied by 80 speakers and 60 bands (*Manchester Evening Chronicle*, 1908). A local newspaper wryly commented that many of the estimated attendees were not motivated by strong feelings about temperance but regarded any political march 'as a relief from the drab monotony of city life' (Editorial, *Manchester Evening Chronicle*, 1908). While this is difficult to prove, it does indicate the difficulty of establishing the motivations of political demonstration attendees and the fact that parks were attractive to a variety of people for a diversity of reasons, some of which cannot be imputed with any certainty. It is possible that a large crowd and the sound of music could function as an attractive element in itself and attendance at such events did not necessarily equate to support for their aims.

A freely available spectacle in a public space, such as a market or a band in a public park, was a viable form of entertainment at a time when recreation was often the preserve of the few. Davies has noted the attraction of open-air markets in Manchester and Salford as a form of popular spectacle for the working classes in the nineteenth and twentieth centuries (Davies, 1988). The variety of spectacles offered by the park was many—music, demonstrations and meetings, sporting activities, horticultural displays—and reinforced the idea of the park as a space to which its visitors had a right. Parks were places where visitors could learn about not just the development of a popular and morally acceptable taste in the arts but also broader concerns, such as citizenship. Public spaces signified arenas where one could not just 'see and be seen' but where the citizens of the future could begin to appreciate their own role in the city and their rights to it—a form of civic socialisation.

Some parks, such as London's Hyde Park, became synonymous with the holding of political meetings and protests, from the Reform League campaign for votes in 1866 to the gathering of the Jarrow marchers in 1936 and its provision of Speakers' Corner, a site for the individual expression of free speech. Parks thus functioned as polemical landscapes of protest and dissent (McKay, 2011). This can be viewed in two ways—as a direct example of the contested nature of public space and a challenge to social control or as a covert mechanism for allowing a modicum of dissent while resisting any real social or political change. The use of urban parks for political purposes was a direct consequence of their public nature. Their very openness was an invitation to gather and exchange views. Yet, as we have seen there were occasions when such meetings could be prevented or contained as the civic authorities struggled to balance control with the principle of general access.

More controversy ensued in 1936 with the granting of permission to the British Union of Fascists (BUF) to hold meetings in British parks. While some concern was expressed about the granting of such permission in Manchester, the chairman of the Parks and Cemeteries Committee, Miles Mitchell, argued that the BUF should be entitled to the same facilities as other groups (*Manchester Guardian*, 1936). Some councillors objected to the use of public parks for such parades, arguing that 'a playground set aside for children . . . (should) not be turned into a battleground' (*Manchester Evening News*, 1936). The meeting was permitted to go ahead but the BUF was not allowed to parade in uniform, which was considered to be inflammatory. The *Manchester Evening News* pointed out that a similar tactic in Germany had had the effect of granting more publicity to the Nazis who paraded in fancy dress instead of uniform (1936). The allure of fascism was temporary, however. Only 300 people attended a BUF meeting in Sheil Park, Liverpool, in February 1938 and a further meeting in March was abandoned when no audience arrived (LA, Parks and Gardens Committee minutes, Volume 35, p. 177 and p. 212).

By the 1930s, ideas about political uses of public space were beginning to relax. This was less a sign of a more tolerant democracy than a consequence of increasing public agitation about access and the right to roam movement, exemplified in the mass trespass on Kinder Scout in Derbyshire in 1932 (Rothman, 1982). Such actions ultimately culminated in the Access to Mountains Act, introduced in 1939 as part of a wider initiative to enable public access to private land (Henry, 2001). While the use of parks for political purposes was an intermittent feature, it often brought to the surface simmering tensions between those who used the parks for the purposes of recreation and those who envisaged a more proactive use for the space. As such, these meetings were just one opportunity among many for societal norms and rules to be violated and contested.

Crime, Violence and Sexual Activity in the Urban Park

Far from being a safe and peaceful haven from the city, the urban park was frequently a site for violent sexual and criminal activity. This ranged from theft of plants, flowers and bulbs to vandalism of seats and sporting equipment and more serious crimes, such as rape, assault and even murder. The design and layout of many parks actually worked to facilitate these behaviours, with their secluded corners and unsupervised spots, which enabled the opportunistic criminal. Actual data on the amounts and types of crimes that occurred in public parks is difficult to obtain, inconsistent and often of a very general nature. Newspaper reports and some police and watch committee statistics give a partial picture of these spaces as criminal environments.

The fact that crimes of many kinds were a regular occurrence in public parks represents an aspect of their incongruity as 'natural' spaces in the unnatural, constructed space of the city. Parks were originally conceived of in opposition to the urban landscape. Much of this stemmed from a tradition of romantic anti-urbanism that privileged rural spaces and rural values over the urban (Marne, 2001). They were supposed to improve not only physical health but also moral health. This was to be achieved through the restorative power of nature, much of which characterised descriptions of the early parks.

Constructing natural environments was not a new undertaking—skilled landscapers, such as 'Capability' Brown, Humphrey Repton and Joseph Paxton, had all created apparently natural gardens for members of the aristocracy. Indeed, Paxton also designed some of Britain's earliest public parks, such as Birkenhead Park and Prince's Park, Liverpool. The ability to create an artificial landscape that appeared natural to the untrained eye was highly prized during the Victorian period and we can find continuities well into the twentieth century. When Manchester accepted the offer of Ernest and Shena Simon of Wythenshawe Park in 1926, the parks superintendent's report on the new park cautioned against interfering with 'the simple and natural beauty of the grounds' (MA, Parks and Cemeteries Committee minutes, Volume 44, p. 43).

Nature was assumed to exert a calming influence on the park visitor. The Official Handbook to Liverpool parks commented that 'it is amid the quiet and restfulness of these sylvan places that the jaded toiler can find real solace and comfort' (O'Mahony, 1934, p. 5). While much of this comment took the form of municipal rhetoric and self-promotion, the emphasis on the naturalness of the park environment was meant as a contrast to the unclean (physically and morally) city that surrounded it. However, it was all too easy for this less pleasant aspect of urban life to make its way into the park and to provide a direct challenge to its moral authority.

Among the main objectionable parks users were the unclean and the vagrant. Manchester's parks bye-laws specifically provided for the removal of 'a person who is offensively dirty' (MA, Parks and Cemeteries Committee minutes, Volume 32, p. 44). There is evidence to suggest that both art galleries and public libraries also suffered from the regular intrusion of vagrancy. It is clear that dirtiness was also considered offensive by other park visitors who actively lobbied the parks authorities to remove vagrants. Henry Coupe, writing to the *Manchester Guardian*, claimed that churchgoers like himself had to 'pick their way . . . to their places of worship' through parkside streets past 'men and women whose cleanliness . . . (was) an open question' (Letters, *Manchester Guardian*, 1908). Park visitors had their own ideas about the right kind of park user and sought to ensure that this was protected. Acts of trespass in parks after their closing time by the homeless and the vagrant were punishable by law. Two men were sentenced to one month's imprisonment with hard labour for sleeping rough in the bowling green toilets in Sefton Park, Liverpool in 1922 (LA, Parks and Gardens Committee minutes, Volume 30, p. 224).

The fact that some park visitors felt that the park was attracting the 'wrong type' of visitor suggests that the parks were not people's parks in the true meaning of the term—parks were contested spaces in which people confronted each other without the possibility of the more delineated demarcation prevalent in other urban spaces, all of which was less easy to establish in the open space of a park. These protests are also indicative of an attempt by some park visitors to encourage or impose their values on others. Historians such as Wyborn have argued that it was the city council that tried to do this, but it is clear from this evidence that park visitors themselves did not agree what constituted a municipal park, how it was to be used or by whom (Wyborn, 1995). Dreher has pointed out the impossibility for parks authorities of enforcing either physical or moral purity in public spaces such as municipal parks (Dreher, 1997).

The nature of many of the crimes that occurred in Britain's public parks reflects their accessibility, openness and attractiveness to a wide range of people, especially children. The presence of children in public parks is emphasised by the frequency with which they were targets of criminals while visiting and playing in the parks. The physical layout of the Victorian park often facilitated those with criminal intent towards park users, with trees, shrubs and bushes for concealment and the availability of sheds, toilets and outhouses within which criminal activity might take place. Many children went to parks alone or in the company of other children and without a supervising adult. A 3-year-old girl died when she fell from a swing at Gaskell Street Recreation Ground in Manchester in 1894. She had been in the company of an 8-year-old girl at the time (MA, Parks and Cemeteries Committee minutes, Volume 14, p. 211).

Sexual acts in public places were a direct challenge to the idea that the natural environment of the park would encourage respectable behaviour. The continual attachment of the epithet 'natural' to public parks was intended to enable the extension of social control in these spaces and to produce a form of moral authority to complement civic authority embodied in the park-keepers and their bylaws. The moral authority of nature and the rhetoric of the natural have a long history in which nature has been prefigured as a type of blank canvas on which diverse ideals can be projected (Daston and Vidal, 2004). The temporal zoning of public parks enabled people to visit by day with the intention of finding a sexual partner for night-time activity—this applied to both heterosexuals and homosexuals (Houlbrook, 2005).

Reports of indecent exposure and paedophile activity regularly occurred. Salford's parks superintendent commented that cases of indecent exposure (usually to women or children) were 'becoming very frequent' and that when apprehended, the culprits were often defended by members of the public who compelled the park-keepers to let them go (SA, Parks Superintendent Report Book, Volume 7, p. 260). John Crawford, aged 43 and of no fixed abode, was sentenced to 12 months' imprisonment with hard labour for indecently exposing himself to little girls in Liverpool's Wavertree Park in 1922. He had a long list of convictions in various parts of Britain (LA, Parks and Gardens Committee minutes, Volume 30, p. 177).

In West Park, Hull, in 1889, Daniel Fenton was found in a shed with a 5-year-old girl on his knee (*Hull Daily Mail*, 1889). He later enticed the child behind some shrubs in the park and attempted to assault her. A similar incident in the same park in 1896 caused the *Hull Daily Mail* to remark that West Park was acquiring an 'unenviable notoriety' (*Hull Daily Mail*, 1896).

Paedophiles and paedophile activity were a problem in many British parks, and it was not uncommon for the same offenders to be apprehended on more than one occasion. Arthur Gall admitted 16 charges of indecent assault and attempted sodomy over a 12-month period in Newsham Park, Liverpool, in 1937 (LA, Parks and Gardens Committee minutes, Volume 34, p. 520). He was sentenced to a year's hard labour. Other, younger offenders received different punishments—a 14-year-old defendant, Harold Chegwin, was sentenced to three years at an approved school and six strokes of the birch for an indecent assault on a 6-year-old girl in Walton Hall Park, Liverpool, in 1938 (LA, Parks and Gardens Committee minutes, Volume 35, p. 363).

Peniston has suggested that paedophiles had little privacy at home, which drove them into public spaces, such as parks (Peniston, 2004). One Salford offender was described by the parks superintendent as a 'regular', suggesting that men with sexual desires for children frequently visited parks due to the availability of their prey and the opportunities for such assault. Far from being peaceful havens from the problems and threats of urban life, public parks often replicated the levels of crime and violence found outside of their gates. The hopes of many of their Victorian founders that people would undergo behaviour modification while enjoying their leisure time in the park were not realised. Parks were as attractive to criminals as they were to other visitors, and the frequency of criminal activity in parks, coupled with the regular reporting of such events in the local newspapers, created the impression that urban parks were sites of conflict, threat and vulnerability.

Couples of both sexes were frequently found engaging in sexual activity in public parks. A man and a woman were found together in the men's toilets in Peel Park, Salford, in 1897. They were sentenced to seven days' hard labour (SA, Parks Superintendent Report Book, Volume 4, p. 26). In Liverpool's Prince's Park, a man and a woman were fined 40 shillings or one month's imprisonment for indecent behaviour in 1923 (LA, Parks and Gardens Committee minutes, Volume 30, p. 376). Women were frequently the focus of unwanted male attention. A complaint relating to Sefton Park in Liverpool alleged that 'men solicit girls in a most persistent and indecent manner, often following them right through the park' (LA, Parks and Gardens Committee minutes, Volume 26, pp. 265–266).

Homosexual acts in public parks were often reported also. In 1936, in Newsham Park, Liverpool, two men received three years' penal servitude and 15 months' hard labour for 'unnatural offences' in a lavatory in the park (LA, Parks and Gardens Committee minutes, Volume 34, p. 311). Where a defendant was substantially older, he received a harsher sentence than a younger man. A 64-year-old man, Thomas Evans of Higher Broughton, Salford, was sentenced to six months' hard labour for acting indecently in a lavatory at David Lewis Recreation ground in Salford in 1931, while his co-defendant, John Jarvis, 19 years old, was bound over in the sum of £10 and to be of good behaviour for two years (SA, Parks Superintendent Report Book, Volume 10, pp. 42–43). Similarly, a man 'of about 38' and 'of the seafaring class' was convicted of committing an indecency with a youth of 16 in Ordsall Park, Salford, in 1929. The older man got nine months' imprisonment with hard labour, while the younger man was discharged (SA, Parks Superintendent Report Book, Volume 9, p. 202).

The 1885 Criminal Law Amendment Act made any act of gross indecency (homosexuality) a crime, regardless of whether it was committed in public or in private. The Labouchere Amendment to the Act, tabled by the Radical MP of that name, explicitly suggested a term of two years with or without hard labour for anyone convicted of such acts (Dellamora, 1990). There was no law preventing lesbian activity. The existence of such 'unnatural' behaviours as homosexuality in the 'natural' environment of the public park raises questions about the success of its moral authority. While it is true that human beings often seek to override or resist any form of authority, be it moral or otherwise, in the name of sexual desire, the prevalence of such acts is an indicator of the continuing difficulties of exerting all but the most basic forms of authority over such complex and multifarious spaces (Daston and Vidal, 2004). It is in the use of parks for sexual behaviours that we find the most direct challenge to the moral authority of the space and, thereby, the most profound failure of the social control paradigm.

Moral arbiters made frequent and public calls for such behaviours to be banished from urban parks. Some blamed the police and parks authorities for the lack of regulation and proper patrolling; others attributed such misbehaviour to the culture of young women whose immodesty was causing the problem. Archdeacon Madden of Liverpool wrote to the Watch Committee about the need to 'train young women in ways of modesty and self-respect' (LA, Parks and Gardens Committee minutes, Volume 26, pp. 265–266). This is as much a reflection of contemporary attitudes towards women's sexuality as passive as of needing to restrain male ardour.

Other crimes were more poignant. In 1915, a young woman was found in Stanley Park in Liverpool in the men's conveniences, trying to dispose of the body of a dead baby girl, to whom she had just given birth (LA, Parks and Gardens Committee minutes, Volume, 28, p. 79). She was hospitalised. Numerous similar instances were reported in parks in other British cities, emphasising the very private nature of what often occurred in a public space. Parks were often the chosen locale for suicide and attempted suicide. Many parks had secluded areas which provided ideal environments for those intent on taking their own lives. There was also a reasonable certainty of being found and, often, of the act itself being prevented. James McNally attempted suicide in Peel Park, Salford, in 1894 by cutting his throat. He was apprehended in the act, charged and released at the Quarter Sessions on condition of his good behaviour for 12 months (*Manchester Courier*, 1894).

Lambert has suggested that such rule-breaking was of a ritualistic nature that was a demonstration of cultural legitimacy (Lambert, 2007). However, designating such behaviour as a ritual implies an element of logic—something planned for and regularly occurring. In the case of the criminal behaviours outlined earlier, much of it seems more opportunistic and transient than ritualistic. While it would be innocuous to claim that the apparently naturally occurring space of the park elicited a correspondingly 'unnatural' response in some users, the attempt to exert the moral authority of the landscape does seem to have engendered the opposite response in some cases.

There is little doubt that nature and the natural appearance of the park were among its most prized attributes. It is entirely possible that these were more highly prized by the civic authorities and by some visitors than others. With so little access to the thoughts and opinions of most park users (especially of the poorer class), some speculation is unavoidable. There were inconsistencies in the recordings of criminal behaviour in public parks, especially of sexual crimes. Some parks superintendents kept diligent and complete records of the names, ages and addresses of those apprehended (Salford, Liverpool); others did not (Preston, Leeds) or did so inconsistently (Manchester).

The impact of the seemingly natural park environment on its users cannot be overestimated. Like the aristocratic estates before them, public parks were systematically designed to leave a particular impression on those who visited them. This pertained as much to a moral impact as a physical one. Nature was to be harnessed for positive, didactic effect—to give pleasure in the form of colourful planting schemes, to teach to inspire and to engender wholesome civic virtues. But pleasure could also have a darker meaning and the search for 'unnatural' pleasures presented those entrusted with administering these parks with their gravest task.

Conclusion

It is tempting to approach the study of the social uses of public parks as a battle for supremacy between the civic authorities and the lawless park visitors. To do so would be to omit the complexities of these relationships, many of which were based more on negotiation and compromise than outright conflict. The evidence of the extensive use of public parks for criminal behaviour and sexual activity is a reminder of the darker sides of such spaces—the opportunities for concealment facilitated by their design and the attractiveness of parks for children meant that the spaces often unwittingly conspired to provide an abundance of possibilities to evade the moral authority supposedly exerted here.

Early efforts to make regular use of public parks for political meetings and demonstrations were becoming less significant as the twentieth century proceeded. The emphasis shifted from these organised demonstrations to more charitable, fund-raising functions, often closely connected to civic life and to more use of parks for the purposes of popular recreation and entertainment. By the 1930s, public parks faced more competition from dance halls and cinemas and their appeal was no longer as obvious to a population more sophisticated and discerning in terms of its leisure tastes. Recreational lives were more commonly being spent in city centres (most parks were at a distance from the heart of the city) or at seaside resorts on the coast.

Local authorities too had less money to invest in the larger parks as they turned their attention to acquiring smaller spaces in more congested and deprived areas to provide basic playgrounds for children. Many of these recreation grounds had little or no green space but were concreted over and had crude facilities, such as sand gardens and swings. They were intended to fulfil a simple function—that of keeping children away from traffic-heavy roads and of providing some physical activity, however mundane. This could be interpreted as a tacit acknowledgement that the larger flagship public parks were not successful in attracting poorer citizens as both their distance from deprived areas and paid-for amenities were a deterrent to the poor. All of this diverse activity raises questions about the success of public parks as 'people's parks'. Not all of these spaces were used equally and by all classes. Their contested nature demonstrates that public space remained open to interpretation by their uses and that they were the sites of many different kinds of unanticipated and disruptive behaviours. This placed increased pressure on those charged with managing and administering these spaces, which forms the subject of the following chapter.

References

Attempted Suicide in Peel Park (1894, 30 August). *Manchester Courier*, p. 7.

Bailey, P. (1998). *Popular Culture and Performance in the Victorian City*. Cambridge: Cambridge University Press.

Beveridge, C. and Hoffman, C. (Eds.) (1997). *The Papers of Frederick Law Olmsted, Volume 1: Writings on Public Parks, Parkways and Park Systems*. Baltimore and London: Johns Hopkins University Press.

Birchall, J. (2006). 'The Carnival Revels of Manchester's Vagabonds': Young Working-Class Women and Monkey Parades in the 1870s. *Women's History Review*, 15(2), 229–252.

Burt, S. (2000). *An Illustrated History of Roundhay Park*. Leeds: S. Burt.

Charge Under Criminal Law Amendment Act (1889, 26 July). *Hull Daily Mail*, p. 4.

Correspondence on the Botanic Gardens (1892, 5 May). *The Belfast Newsletter*, p. 6.

Daston, L. and Vidal, F. (2004). Doing What Comes Naturally. In L. Daston and F. Vidal (Eds.). *The Moral Authority of Nature* (pp. 1–20). Chicago and London: University of Chicago Press.

Davies, A. (1988). Saturday Night Markets in Manchester and Salford 1840–1939. *Manchester Region History Review*, 1(2), 3–12.

Dellamora, R. (1990). *Masculine Desire: The Sexual Politics of Victorian Aestheticism*. Chapel Hill and London: University of North Carolina Press.

Dingle, A. E. (1980). *The Campaign for Prohibition in Victorian England*. London: Croom Helm.

Dreher, N. (1997). The Virtuous and the Verminous: Turn-of-the-Century Moral Panics in London's Public Parks. *Albion: A Quarterly Journal Concerned With British Studies*, 29(2), 246–267.

Early Morning Outrage (1913, 12 November). *Manchester Guardian*, p. 16.

Editorial (1896, 4 June). *The Manchester Courier*, p. 5.

Editorial (1908, 29 June), *Manchester Evening Chronicle*, p. 2.

Eighty Speakers (1908, 26 June). *Manchester Evening Chronicle*, p. 4.

Etheridge, S. (2014). *Slate Grey Rain and Polished Euphoniums: Southern Pennine Brass Bands, the Working Class and the North c.1840–1914* (Unpublished PhD Thesis), University of Huddersfield, Huddersfield.

Fascists May Parade, But Not in Uniform (1936, 8 October). *Manchester Guardian*, p. 11.

Fascists Must Doff Their Uniforms for Park Meeting (1936, 7 October). *Manchester Evening News*, p. 8.

Griffin, C. (2008). 'Cut Down by Some Cowardly Miscreants': Plant Maiming or the Malicious Cutting of Flora as an Act of Protest in Eighteenth and Nineteenth Century England. *Rural History*, 19(1), 29–54.

Gurney, P. (1997). The Politics of Public Space in Manchester 1896–1919. *Manchester Region History Review*, 11, 12–23.

Henry, I. (2001). *The Politics of Leisure Policy* (2nd Ed.). Basingstoke: Palgrave Macmillan.

Hill, K. (2001). 'Roughs of Both Sexes': The Working Class in Victorian Museums and Art Galleries. In S. Gunn and R. Morris (Eds.). *Identities in Space:*

Contested Terrains in the Western City Since 1850 (pp. 190–103). Aldershot: Ashgate.

Houlbrook, M. (2005). *Queer London: Perils and Pleasures in the Sexual Metropolis 1918–1957*. Chicago and London: University of Chicago Press.

Hunt, K. (2012). The Local and the Everyday: Interwar Women's Politics. *The British Historian*, 42(4), 266–279.

Jones, G. S. (1983). *Languages of Class: Studies in English Working Class History 1832–1982*. Cambridge: Cambridge University Press.

Joyce, P. (1991). *Visions of the People: Industrial England and the Question of Class 1840–1914*. Cambridge: Cambridge University Press.

Lambert, D. (2007). Rituals of Transgression in Public Parks in Britain, 1846 to the Present. In M. Conran (Ed.). *Performance and Appropriation: Profane Rituals in Gardens and Landscapes* (pp. 195–210). Boston: Harvard University Press.

Lancashire Record Office, Preston, Parks and Recreation Grounds Committee Minutes, CBBU/39.

Langhamer, C. (2005). Leisure, Pleasure and Courtship: Young Women in England 1920–1960. In M. J. Maynes, B. Soland and C. Benninghaus (Eds.). *Secret Gardens, Satanic Mills: Placing Girls in European History 1750–1960* (pp. 269–283). Bloomington and Indianapolis: Indianapolis University Press.

Lasdun, S. (1991). *The English Park: Royal, Private and Public*. London: Andre Deutsch.

Layton-Jones, K. and Lee, R. (2008). *Places of Health and Amusement: Liverpool's Historic Parks and Gardens*. Swindon: English Heritage.

Letters (1908, 22 July). *Manchester Guardian*, p. 5.

Liverpool Archives, Liverpool, Parks and Gardens Committee Minute Books, 352/MIN/PAR/1/.

Manchester Archives and Local Studies, Manchester, Parks and Cemeteries Committee Minute Books, GB127.Council Minutes/Parks and Cemeteries/1-53.

Marne, P. (2001). Whose Public Space Was It Anyway? Class, Gender and Ethnicity in the Creation of Sefton and Stanley Parks. *Social and Cultural Geography*, 2(4), 421–443.

McCrone, K. (1988). *Sport and the Physical Emancipation of English Women 1870–1914*. London: Routledge.

McKay, G. (2011). *Radical Gardening: Politics, Idealism and Rebellion in the Garden*. London: Frances Lincoln.

Mitchell, H. (1968). *The Hard Way Up: The Autobiography of Hannah Mitchell, Suffragette and Rebel*. London: Faber and Faber.

Neal, F. (1988). *Sectarian Violence: The Liverpool Experience 1819–1914*. Manchester: Manchester University Press.

O'Mahony, M. (1934). *Official Handbook: The Parks, Gardens and Recreation Grounds of the City of Liverpool*. Liverpool: Liverpool City Council.

The Opening of Horton Park, Bradford (1878, 1 June). *The Leeds Times*, p. 2.

Opening of the Public Parks (1846, 28 August). *The Manchester Times*, p. 6.

Park Pests (1896, 24 June). *Hull Daily Mail*, p. 4.

Penistone, W. (2004). *Pederasts and Others: Urban Culture and Sexual Identity in Nineteenth Century Paris*. New York: Harrington Park Press.

Pettigrew, A. A. (1926–1932). *The Public Parks and Recreation Grounds of Cardiff*. 6 Volumes. Unpublished.

Public Opening of the Sefton Park Playground and Gymnasium (1862, 12 March). *Liverpool Mercury*, p. 5.

Pugh, M. (1994). *State and Society: British Political and Social History 1870–1992*. New York: Routledge.

Rodrick, A. (2004). *Self-Help and Civic Culture: Citizenship in Victorian Birmingham*. Aldershot: Ashgate.

Rosen, A. (1974). *Rise Up, Women! The Militant Campaign of the Women's Social and Political Union 1903–1914*. London: Routledge and Kegan Paul.

Rothman, B. (1982). *The 1932 Kinder Scout Trespass*. Altrincham: Willow Publishing.

Ruff, A. (2000). *The Biography of Philips Park, Manchester 1846–1996*. Manchester: University of Manchester School of Planning and Landscape, Occasional Paper, 56).

Salford Archives, Salford. Parks Superintendent Report Books. L/CS/DR4/1-13.

Steinbach, S. (2012). *Understanding the Victorians: Politics, Culture and Society in Nineteenth Century Britain*. London and New York: Routledge.

Stevenson, J. (1984). *British Society 1914–1945*. London: Penguin.

Taylor, D. (1999). Central Park as a Model for Social Control: Urban Parks, Social Class and Leisure Behaviour in Nineteenth Century America. *Journal of Leisure Research*, 31(4), 420–477.

Thompson, E. P. (1963). *The Making of the English Working Class*. London: Penguin.

Tonight by Tempus (1936, 7 October). *Manchester Evening News*, p. 8.

West Yorkshire Archives Service, Leeds. Parks Committee Minute Books. LLC47/1/.

The Women's Demonstration (1908, 20 July). *Manchester Evening News*, p. 3.

Women's Big Day (1908, 20 July). *Manchester Evening Chronicle*, p. 3.

Wyborn, T. (1995). Parks for the People: The Development of Public Parks in Manchester. *Manchester Region History Review*, 9, 3–14.

4   Public Parks as Employers

Introduction

Writing in 1933, the parks superintendent of Manchester, W. W. Pettigrew, observed that

> it is generally admitted by those best fitted to judge that a trained lawyer makes the most efficient town clerk. For the same underlying reason it must be agreed that the trained and experienced horticulturalist makes the most successful parks superintendent.
>
> (W. Pettigrew, 1937, p. 189)

That he should believe this to be so is not surprising as he was himself from a horticultural background. This statement, however, also hints at another aspect of this most challenging municipal position—the ability to combine a variety of different skills for a public purpose. As Pettigrew also pointed out, the job title itself often varied from one town or city to another—Parks Superintendent, Chief Superintendent or Director of Parks. After the formation of the Institute of Park Administration in the 1920s, most began to use the title Parks Superintendent, although this was by no means unanimous.

This role was arguably the most significant and influential role in the parks department (whose own title altered accordingly) and one that imposed many demands on the individual carrying it out. The parks superintendent took responsibility for the enactment of municipal policy but regularly found himself torn between the demands of his employer and the park users. The legacy of the superintendent is a meticulous record of achievements in bringing the delights of green space to a maximum number of urban dwellers. Parks departments in general were one of the largest in terms of employees in any town, city or borough council. Thus, they are worthy of study in their own right as training grounds, vehicles of public policy and mechanisms for managing and administering often-competing needs for public recreation and leisure.

Managing and Administering the Public Park

The evolution of the role of the parks superintendent was gradual, and the individuals given this title often emerged out of the general pool of park superintendents who each had charge of a particular park. Prior to the existence of the position, most cities relied on borough engineers and city surveyors to source open space for parks and, in many cases, to be involved in their design. This was the case in Liverpool, when the borough engineer, James Newlands, and the district surveyor, John Weighman, were tasked by the Improvement Committee with investigating the opportunities for land for public parks to the north of the city in 1864 (LA, Parks and Gardens Committee minutes, Volume 4, p. 339). In some instances, staff were also involved in the design and layout of the parks— Weighman's successor in Liverpool, E. J. Robson, was involved in the laying out of Stanley Park.

Similarly, in Cardiff, the borough engineer initially managed all of the parks and was responsible for finding new sites, preparing plans and laying out parks (A. Pettigrew, 1932, Volume 2, p. 63). In 1891, the decision was taken to formally appoint a head gardener and three assistants to manage the laying out of Roath Park in particular. After advertising the jobs, William Wallace Pettigrew was appointed as head gardener at a salary of £120 per annum to work under the direction of the borough engineer (Figure 4.1). As was later pointed out, in practice, Pettigrew mostly reported directly to the Parks Committee and it communicated straight to him without going through the borough engineer at all (A. Pettigrew, 1929, Volume 2, p. 63).

In January of 1896, Pettigrew was given the title of General Superintendent of Public Parks and Open Spaces and was made responsible directly to the Parks Committee. Parks had become an independent department in their own right (A. Pettigrew, 1929, Volume 2, p. 64). Pettigrew was allocated an official residence in Roath Park from 1897 (A. Pettigrew, 1929, Volume 2, p. 64). Gestures such as this became common practice in inducing the best candidates for these posts. This was a reflection of the fact that such local government officials were 'expected to be creative figures, not simply agents obeying orders and executing routine work' (Waller, 1983, p. 282). This also signalled the transformation of municipal authorities into 'complex bureaucracies' that needed professional staff to meet the demands of the growing urban environment (Chandler, 2007, p. 84).

As cities grew and added to their stock of open, green space, the need arose for a more formal management and policy development and implementation specialist. Often, these men were drawn from the pool of already-existing parks superintendents, but by the late nineteenth century, these had emerged as high-profile and competitive roles. The job title Parks Superintendent was replaced more usually by Director of Parks

Figure 4.1 William Wallace Pettigrew, General Superintendent of Parks, Manchester.
Source: © Tim Pettigrew collection.

during the 1920s and 1930s (1937 in Manchester, Manchester Archives, Parks and Cemeteries Committee minutes, Volume 52, p. 42). This decision was deemed necessary to bring Manchester into line with established practice in other cities and reflects a trend of increasing managerialism in Britain at that time. Sometimes, the title was expanded beyond parks to include other related responsibilities. In Leeds, the title of Parks Superintendent was altered to Director of Parks, Cemeteries and Allotments in 1924 (WYAS, Leeds Parks Committee minutes, Volume 3, p. 125).

In Manchester in 1893, such was the attention being paid to the pressures of the role that the Parks and Cemeteries Committee established a special subcommittee entitled 're. the duties etc. of the General Superintendent of Parks' to examine the post more thoroughly (MA, Parks and Cemeteries Committee minutes, Volume 14, p. 4). In December of the previous year, it had been proposed to increase the parks superintendent's salary from £156 per year to £250, a substantial rise. This decision was altered in January 1893 to £200 and it is possible that it was agreed to re-examine the post as a result of this decision (MA, Parks and

Cemeteries Committee minutes, Volume 13, p. 84). One of the decisions of the subcommittee was to provide a house free of rent for the general superintendent (MA, Parks and Cemeteries Committee minutes, Volume 14, p. 4)—this was a standard part of the emoluments for this post in other cities.

By 1893, the parks superintendent in Hull, Edward Peak, wrote to the Parks and Recreation Grounds Committee to ask for an increase in his salary. Having been employed in post for 33 years and being in charge of three parks, 11 disused burial and sanatorium grounds and 2,850 street trees, he wrote that 'I believe that I am the worst paid superintendent with similar duties in the country' (HHC, Parks and Burials Committee minutes, Volume 6, pp. 71–72). He was granted an increase from £130 to £150 annually. This reflects the fact that there were varying practices around the country as regards the salaries for these very responsible posts and those occupying them often had to plead their own cases on their merits.

Allied to these concerns were the increasing administrative pressures on parks departments' work after World War I. Salford's parks superintendent reported a 'great increase of office work owing to the various reports and stats continually required by the Town Clerk's department and others' in 1922 (SA, Parks Superintendent Report Books, Volume 8, p. 31). This was undoubtedly due to wider and more general expansions in the administration of local government and the rise of the civil servant class (Garrard, 1995). A new profession of administrative and other bureaucrats placed demands on all local government departments of the kinds outlined by Wilsher. He had had only one clerical assistant in the Salford Parks department since 1908. This was eventually augmented by an additional typist appointed in 1927 to manage the extra work occasioned by increased provision for games and music in Salford's parks (SA, Parks Superintendent Report Books, Volume 9, p. 7).

This situation placed increasing administrative pressures on parks superintendents and removed them from the task of strategically managing parks policy. This was alleviated to some extent in Salford in 1923 by the appointment of an assistant parks superintendent, whose job it was to concentrate on the development of existing parks in the city, leaving the parks superintendent to focus on new work, office work and general management (SA, Parks Superintendent Report Books, Volume 8, p. 35). An administrative assistant had been granted to Manchester's general superintendent in 1903 in order to support him in his duties and as a reflection of the increasing administrative pressures (MA, Parks and Cemeteries Committee minutes, Volume 23, p. 84).

The role of parks superintendent was a key one in terms of the development and delivery of a strategy for the continual enhancement of a town or city's parks (which often included cemeteries as well). The parks superintendent could be an influential figure and one whose work brought with it some local renown. When William Wallace Pettigrew arrived in Manchester from Cardiff in 1915 to take up the post of parks superintendent, his appointment was welcomed by one of the local newspapers as that of a foster father to the municipal parks. Somewhat fancifully, the *Manchester Courier* asked whether the new appointee might treat his children to 'the pretty dresses of his predecessor' or whether he would develop the parks in a new direction, suggesting that some innovation would be welcome (27 April, 1915, p. 6). Pettigrew had trained at the Royal Botanic Gardens in Kew. His father had been the head gardener at Dumfries House and his younger brother Andrew took over from him in Cardiff as parks superintendent on his appointment to Manchester.

The high-profile nature of this post is a testament to its importance in the development of civic amenities for public recreation. Pettigrew took up his post in Manchester at a key time in the evolution of its parks and just as the real implications of World War I for municipal parks were becoming apparent. The importance attached to the post and the level of skills required were often reflected in the salaries offered to these officials. In 1902, the annual salary of Alfred Wilsher in Salford was increased from £225 to £250 (*Manchester Evening News*, 1902). This rate of pay was comparable with the chief audit clerk of the Borough Treasurer's department and the chief clerk of the Health Committee.

Parks superintendents in most cities were also granted additional benefits, such as a dwelling house, coals and gas, free of charge. These posts, when they were advertised, were competitive, with candidates often travelling from a distance for interview. Seven candidates were interviewed for the post of parks superintendent in Leeds in 1923 when the salary was £450 per annum and included a house, coals and light (WYAS, Leeds Parks Committee minutes, Volume 3, p. 76). The following year, the title of the post was widened considerably to Director of Parks, Cemeteries and Allotments (WYAS, Leeds Parks Committee minutes, Volume 3, p. 125), reflecting the continuing expansion of the scope of the role. A similar decision was taken in Manchester to alter the title of the role from General Superintendent of Parks to Director of Parks and Cemeteries in 1937 (MA, Parks and Cemeteries Committee minutes, Volume 52, p. 42). The emphasis on 'Director' indicates the increasing seniority of the job but also an increasingly hierarchical public service ethos, which was pervading many local authorities.

By 1928, Preston City Council was paying its new parks superintendent, Algernon Birkinshaw, £330 per annum as well as providing a house free of charge (LRO, Parks and Baths Committee minutes, Volume 9, p. 392). Not only were all qualified candidates for this post interviewed but also the Parks and Baths Committee chairman visited all of the cities in which they were employed, in order to assess their work prior to interview (LRO, Parks and Baths Committee minutes, Volume 9, p. 368). Ten years later, the salary had increased to £405 with a house free of charge (LRO, Parks and Baths Committee minutes, Volume 10, p. 162).

This suggests a certain level of anxiety to appoint a suitable candidate to this prestigious post and a desire to keep a successful candidate in post. Of course, these committees were also accountable to the ratepayers for decisions about money to be spent on salaries and for verification that value for money was being obtained.

Such were the demands of the job, by the 1920s, that Preston's parks superintendent had a bicycle supplied to him to enable him to cover the distances between the parks he administered efficiently (LRO, Parks and Baths Committee minutes, Volume 9, p. 256). By 1929, the bicycle was upgraded to the supply of a motor car at a cost of £225 (LRO, Parks and Baths Committee minutes, Volume 9, p. 408). By the early 1930s, electric lighting was being installed in the parks superintendent's house in Preston, and his salary had increased to £405 by 1938 (LRO, Parks and Baths Committee minutes, Volume 10, p. 24 and p. 162).

Parks superintendents were responsible for decision-making about departmental parks employee uniforms and for employing, managing and appraising all parks department staff. The evolution of the parks employee uniforms provides an instance of the growing awareness of the need to present a consistent and authoritative public face within these departments. The annual specification and ordering of the uniforms formed part of the parks superintendent's duties as monies had to be approved by the relevant Parks Committee. The style and composition of the uniform varied from one city or borough to another and from year to year. A Liverpool park-keeper's clothing in 1870 consisted of a frock coat, a pair of trousers, a vest, a pair of gloves and a pair of boots (LA, Parks and Gardens Committee minutes, Volume 7, p. 306). By 1881, a park-keeper's uniform was composed of trousers, a body coat, a great coat, gloves, hat with band and lettering, a cape and boots. The following year, leggings were added to the uniform, presumably as an extra layer for winter (LA, Parks and Gardens Committee minutes, Volume 13, p. 18 and p. 242). Not every city was as thorough in clothing its employees—Hull's park-keepers were each allocated a coat, a vest, two pairs of trousers and a waterproof topcoat in 1893 (HHC, Parks and Burials Committee minutes, Volume 6, pp. 21–22).

Parks superintendents were regularly interviewed in the local press about their plans for the future of the parks, and they provided insights for readers about new strategies and decisions. The *Hull Daily Mail* interviewed its new parks superintendent, Harry Bursell Witty, in 1902 about his new strategy for moving flowering plants closer to the parks entrances in order to enhance the visitor's experience on immediately entering the park. The paper noted Witty's 'energy and capacity' and commented that he was the 'right man in the right place' (24 April, 1902, p. 4).

Similarly, Manchester's William Wallace Pettigrew began to write a weekly column for the *Manchester Guardian* from 1926, entitled 'Nature Notes From Manchester Parks', in which he extolled the virtues of the various horticultural initiatives of the parks department. Pettigrew himself outlined the three benefits that he hoped to derive from this regular publicity: to increase public interest in and knowledge of botany, to provide educational value and to encourage the public to help the Parks staff to protect the plants from damage (MA, Parks and Cemeteries Committee minutes, Volume 43, p. 134). The plea to all parks users to collaborate in ensuring the security of the parks' amenities was more than a general concern about social order but also an important mechanism of involving all users in accepting active responsibility for maintaining the space, a key component of responsible citizenship.

Pettigrew also gave public talks about the activities of the parks department and maintained a collection of about 100 slides of various park scenes for this purpose (MA, Parks and Cemeteries Committee minutes, Volume 46, p. 236). This readily available information about horticulture made the subject accessible to all social classes and opened up the possibility of utilising even the smallest domestic gardening space to good advantage. Pettigrew published *Common Sense Gardening* in 1925, which was aimed at novice gardeners. Willes (2015) has commented on the importance of such literature as a source of inspiration and practical advice for working-class gardeners in particular. Having witnessed the displays of flowers in a public park, the domestic gardener could learn similar techniques from the parks superintendent himself.

All of this newspaper comment also provided useful advertising for the parks department and a timely reminder to the readers of the amenities offered. On the occasion of Witty's retirement in 1929, the *Hull Daily Mail* reminded its readers that the 'ramifications of the position have grown immensely' (7 November, 1929, p. 11). On occasion, the pressures of the post had serious consequences for the health of the parks superintendent. The parks superintendent in Hull was granted two weeks of leave to 'undertake a sea voyage for the benefit of his health' in 1889 (HHC, Parks and Burials Committee minutes, Volume 6, p. 45). Liverpool's chief parks superintendent and curator J. J. Guttridge had to take two months off work in 1921 to convalesce from a nervous breakdown (LA, Parks and Gardens Committee minutes, Volume 27, p. 67). In Salford, the parks superintendent was advised by his doctor to take two weeks for 'immediate rest' in 1937 (SA, Parks Superintendent Report Books, Volume 13, p. 45).

While there was no explicit link made between the physical and mental health of the parks superintendents and the professional strain of their jobs, it seems likely that, given that these very experienced men suffered from these illnesses, it was certainly a factor. While they all recovered, these incidents are an indication of the increasing stress experienced by parks superintendents, as the numbers of parks and recreation grounds increased, often without an increase in administrative support. The parks superintendent was a more high-profile figure than previously and this added scrutiny to what had previously been a somewhat lower-profile and rather more manageable role. Increases in the numbers of parks during the early decades of the twentieth century, combined with more, smaller recreation grounds and a more demanding and leisure-conscious public, meant that parks superintendents found themselves under more pressure to manage both parks and staff.

The retirement of a long-standing and effective parks superintendent caused potential problems for many local authorities. In Liverpool in 1932, the parks superintendent was prevailed upon to continue for 12 months past his retirement date at a salary of £1,000 per annum with house, coals and gas (LA, Parks and Gardens Committee minutes, Volume 33, p. 239).

While the post of parks superintendent was a high-profile one, this could occasionally pose problems. In 1930, the parks superintendent in Hull, George Copley, was forced to resign his position after a scandal. Copley had been appointed as parks superintendent only in November 1929, succeeding H. B. Witty, who had held the position for over 40 years. Copley had been investigated by a special subcommittee of the Parks and Burials Committee for irregularities with respect to the sale of municipal goods. While the investigation took place, Copley was suspended from his duties and required to vacate his office in Pearson Park.

The special subcommittee concluded that his salary be reduced by £100 per year but that he be retained as parks superintendent, subject to a severe reprimand (HHC, Hull Parks and Burials Committee minutes, Volume 6, p. 166). A full committee vote to ask for his resignation was lost. On 13 November 1930, Copley offered to resign, subject to three months' notice. A meeting of the full city council decided to ask for his immediate resignation and, if this was refused, that Copley be dismissed immediately.

Copley then resigned with immediate effect (HHC, Parks and Burials Committee minutes, Volume 6, p. 2). An advertisement was issued for his replacement, which resulted in 79 applications, one of which was from Copley (presumably signifying that he felt that he was innocent of the charges). He was not placed on the shortlist and James Roberts, the parks superintendent of Chesterfield, was duly appointed, with a stipulation that he devoted all of his time to the post and not engage either directly or indirectly in any other business or employment (HHC, Parks and Burials Committee minutes, Volume 6, p. 27). This may offer a clue as to the difficulty posed by Copley's alleged behaviour.

The local press took an intense interest in these events. Copley's predicament was one of three inquiries going on at this time in various departments of Hull City Council and one that served to create the impression of a sense of crisis in the local authority. The *Hull Daily Mail* criticised what it termed the 'hush-hush' policy of the city council in respect of these inquiries, which were held in camera and which, it argued, caused more than usual interest and apprehension among the city's ratepayers (20 October 1930, p. 1).

The lingering sense of malfeasance at the city council continued after Copley's resignation with controversy about the relationship between the city engineer and the parks superintendent. This arose from a question raised by some of the candidates interviewed for the post of parks superintendent in December 1930. A clause in the employment agreement for the parks superintendent made reference to the preparation of the estimates in conjunction with the city engineer and the practice of sending wages and time sheets from the parks department to the city treasurer. This was perceived as a potential opportunity for the city engineer to interfere with the work of the parks superintendent. It was not a local controversy, however, with Alderman Stark pointing out that a 'war was being waged in the country between these officials' (*Hull Daily Mail*, 18 December 1930, p. 1). Nevertheless, it ensured that the issue of the role of the parks superintendent remained on the front page of the local newspaper and resulted in the reassurance of the candidates by the chairman of the Parks and Burials Committee that the parks superintendent had entire control of all matters appertaining to the parks proper.

Alderman Stark's comment presents an opportunity to consider the emergence of a professional class of administrators in local government during this period. Garrard (1983) has noted the preponderance of local businessmen on city councils in many British towns and cities in the nineteenth century. This remained the case in cities such as Manchester until the early decades of the twentieth century. But municipal leadership was about more than just elected officials. Indeed, much of the business of local governance was accomplished with the skills and patience of administrative officials, such as town clerks, parks superintendents and borough engineers. Garrard (1995) has described the relationship between the administrators and elected officials as a form of power-sharing based on 'scarce and specialised expertise' (p. 598). This knowledge was enhanced by membership of a growing number of professional organisations, such as the Institute of Park Administration (1926), which afforded opportunities to meet and share skills. While the numbers of businessmen on local councils diminished in the early twentieth century, the bureaucratic class expanded with the appointment of deputies to assist in the expanding work. One of the main responsibilities of administrators such as parks superintendents was the management of the wide-ranging staff employed in the parks.

Parks Employees

The business of staffing Britain's public parks often had modest beginnings. As the numbers of parks and open spaces grew, most towns and cities expanded their parks staff. In Cardiff, for example, the parks department consisted of two gardeners in March 1890, who were responsible for the care and maintenance of five gardens between them (A. Pettigrew,

1929, Volume 2, p. 14). By July 1890, five caretakers augmented these, one for each garden (A. Pettigrew, 1929, Volume 2, p. 14). By 1898, the parks department had 74 workmen on the payroll, including the park-keepers (A. Pettigrew, 1929, Volume 2, p. 64). This number had increased to 77 by 1905 (A. Pettigrew, 1929, Volume 2, p. 64).

The increasing professionalisation of park management and administration in the early twentieth century is illustrated by the gradual relaxation of attempts to control visitor behaviour directly. In general, Victorian parks drew their visitors' attention to the parks regulations principally through prominently displayed noticeboards, whose rules were enforced by park-keepers. As late as 1937, W. W. Pettigrew, the general superintendent of Manchester parks, wrote that such notices were necessary as park visitors needed to know what was allowed and what was not, thus establishing the boundaries of acceptable behaviour (W. Pettigrew, 1937, p. 88). Prior to the official opening of Heaton Park in 1902, the *Manchester Guardian* noted the enforcement of Corporation regulations in the park: 'When it was lent to the public at Whitsuntide by its late owner, one could wander everywhere. Yesterday, there were policemen and keepers almost by the score, who gave the word to "keep off the grass"' (27 June, 1902, p. 6).

This nostalgia for the former aristocratic owner omits the fact that under aristocratic ownership, the park was rarely accessible to the general public. The following day, the *Manchester Guardian* expressed the hope that 'too much zeal will not distinguish the well-meaning custodians of our new park or the citizens who make the journey there may begin to feel that they are being dragooned into revolt' (28 June, 1902, p. 5). The *Manchester City News* continued this theme the following year as it described how people walked on the grass in spite of being asked to refrain by the park-keeper: 'It is curious to note . . . that gentleman always walks on the grass himself, forgetting the moral influence of that action' (25 July, 1903, p. 3).

Contemporary poems satirised and bemoaned what was often perceived as the overly officious regulation of public parks and those who frequented them. *Punch* (p. 64) published the following in 1881:

THE PARKS FOR 'THE PEOPLE!'

(Yes, but what Sort of 'People'?)
Whene'er I take my walks abroad,
How many sots I see,
And though I never speak to them,
They often speak to me.
I do not heed their coarse remarks,
But with their playful curses,
They frighten from our healthful parks
The children and their nurses.

Similarly, the *Manchester Evening News* published a poem written by a schoolboy, Austin Latham, in 1929 (11 January, 1929, p. 7):

PARKS

I don't like parks a little bit
Where grown-up people go and sit
And little boys in woolly gaiters
Walk 'spectably by perambulators
Where writing says: 'keep off the grass'
And little dogs with ladies pass
I don't like parks at all—do you?
With all the things you mustn't do.

However, it is also true that the role of the park-keeper was about much more than mere enforcement of the rules. This was another position that increased in responsibility and complexity at the end of the nineteenth century. The park-keeper was in overall charge of a park and the staff. The role originally combined working and watching (security) but these two were gradually disaggregated from each other (Lambert, 2005). The park-keeper was responsible for opening and closing the park gates at the designated hours and for making sure that the public left the park in good time. They kept order and discipline, which was often problematic when the parks were busy during the school holidays. The park-keeper at Stanley Park in Liverpool remarked that the park was often so overcrowded with school children and 'to a large extent by a destructive and hooligan type' that dealing with wear and tear on the landscape was difficult (LA, Parks and Gardens Committee minutes, Volume 27, p. 191).

Park-keepers were expected to have horticultural skills in terms of deciding how and when to plant and maintain the parks and to have people management skills to manage their staff. Their authority was signalled to their staff and the general public alike by their uniform, which often resembled that of a police officer. Many were sworn in as special constables on appointment, which allowed them to exercise powers of arrest if necessary. The 1872 Parks Regulation Act allowed for park-keepers to be able to act as park constables within the area in which the park was located (Parks Regulation Act, 1872, section 7). The Act also provided for a penalty of £50 or up to six months in prison if a person was found guilty of assaulting a park-keeper (Parks Regulation Act, 1872, section 6). Police constables in the area of a park were also entitled to act as park constables within the environs of the park (Parks Regulation Act, 1872, section 8). The park-keeper's salary reflected the level of management required and the size and prominence of the associated park. In Liverpool in 1868, park-keepers earned 20 shillings a week (LA, Parks and Gardens Committee minutes, Volume 6, p. 231). Some park-keepers employed at larger parks also got a house in that park as well as an allowance of coals and gas.

The amount of time spent on patrolling and watching duties expanded as parks grew in number and popularity with the public. Some local authorities began to explore the possibility of plain-clothes policemen undertaking patrolling instead of the park-keeper. A suggestion by the Hull parks superintendent that this be introduced in 1922 was refused by the chief constable due to cutbacks in the forces numbers by the Geddes Committee (HHC, Parks and Burials Committee minutes, Volume 6, p. 73). The city of Liverpool's Watch Committee proposed replacing park-keepers with regular police altogether in 1918 (LA, Parks and Gardens Committee minutes, Volume 28, pp. 543–544). The city's parks superintendent defended the role of the park-keeper and pointed out that the police were effective only after a crime had taken place, whereas the park-keeper had a more preventative role. He argued that park-keepers were therefore more effective than the police, although there was a problem evidencing this due to the fact that the numbers of warnings and checks on behaviour by the park-keepers were not formally recorded. Watching was often seasonal in nature—more staff were needed in the summer when the parks were busier. In Liverpool, summer park-keepers were often less experienced and did not wear uniforms so, as the parks superintendent commented, 'the public do not heed them as they would a uniformed man' (LA, Parks and Gardens Committee minutes, Volume 27, p. 192).

In spite of attempts both thwarted and successful to regulate and patrol public parks, it is also clear that working in a park for the ordinary labourer and gardener was an arduous job. The work in most parks was heavy, demanding manual labour. It required a high degree of sustained physical effort and often took its toll on the older employees. The daily working life of the parks superintendent and his employees offers a fascinating glimpse into a hitherto unknown and fairly undocumented aspect of park life. Alfred Wilsher, the parks superintendent for Salford from 1894 until 1926, has left a detailed account of this working life in his weekly written reports, which were presented to the Parks Committee (Figure 4.2). These reports document the weekly activities of his men, the plans underway for various events and park festivals, crimes committed in the parks and such details as levels of sickness absence from the city's parks.

Wilsher outlined in detail the usual activities of the men—digging, weeding, planting flowers, shrubs and trees, pruning, maintaining and enlarging the flower beds over a considerable number of years (SA, Parks Superintendent Report Books, Volumes 1–13). This work went on in the parks year round, as can be seen in Figure 4.3. The parks superintendent in Hull detailed the work in East Park undertaken by the labourers during the winter of 1896—six acres of shrubberies were cleaned and dug over, 108 flower beds were edged around, dug over and manured, gravel walks were repaired, the cricket grounds were re-sod and 684 avenue trees were pruned (HHC, Parks and Burials Committee minutes, Volume

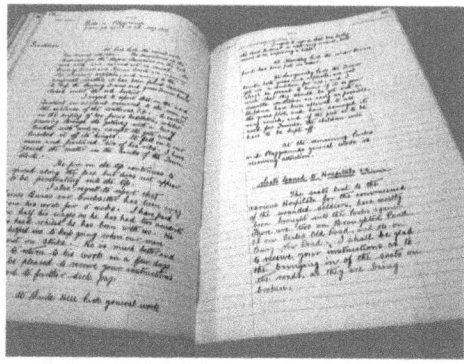

Figure 4.2 Parks Superintendent Report Book, Salford.

Source: Photo by and © Carole A. O'Reilly.

6, pp. 81–82). The physical nature of the work is emphasised by the regular reporting of those incapacitated and unable to work. These were logged on a weekly basis and, if the illness proved long-term, Salford's parks superintendent Wilsher inquired of the committee as to what level of wages (if any) should continue to be paid (SA, Parks Superintendent Report Books, Volume 4, p. 32 and p. 132). The level of wage that was paid was generally related to the length of service of the individual concerned. Those who had worked for the parks for many years (as many as 25 to 30 years in some instances) were more likely to be paid most if not all of their wages when they were ill.

Some parks workers were members of a 'sick club', a type of friendly society that encouraged self-help for those who could afford it and which paid sickness and death benefits for members (Hopkins, 1979, p. 142). In 1906, 28 employees of the Salford Parks department were in a sick club, while 47 were not (SA, Parks Superintendent Report Books, Volume 5, inside front cover). The comparatively small numbers in the sick club indicates that most employees could not afford this type of protection and were, therefore, reliant on the goodwill of the Parks Committee if and when they suffered an illness or injury. Kidd has calculated that, by the second half of the nineteenth century, more than one in three adult males were able to make private provision for themselves and their families through membership of a friendly society (Kidd, 2002). These friendly societies were part of an important network of mutual aid schemes that developed during the nineteenth century and were often aligned to specific occupations (Fraser, 2009). The average contribution varied from 4 to 9 pence a week (Hanson in Hartwell, 1972).

The physical nature of the work also took its toll on the employees. In Liverpool in 1885, it was decided by the Parks and Gardens Committee to reduce the wages of all parks labourers who reached the age of 65 (LA,

Figure 4.3 Gardening staff at Peel Park, Salford.

Source: © The Francis Frith Collection.

Parks and Gardens Committee minute books, Volume 12, p. 447). Two labourers from Wavertree Park complained about this decision on the basis that although they were 65, they were 'capable of doing a fair day's work' (LA, Parks and Gardens Committee minute books, Volume 12, p. 418). The parks superintendent noted that they were both honest and industrious men and the original decision was rescinded (LA, Parks and Gardens Committee minute books, Volume 12, p. 418).

Dealing with ageing employees was a constant problem for the parks service. There was often a reluctance to sideline or terminate altogether their employment after reaching a certain age but a pragmatic approach was often needed. Hull's parks superintendent reported in 1889 that two of the three labourers at Pearson Park were struggling to perform their jobs adequately. He remarked that he would be 'sorry to do or say anything that would tend to deprive the men of their employment' (HHC, Parks and Burials Committee minutes, Volume 6, p. 12). One of the two had been employed at Pearson Park since its formation and the other for 18 years. In view of their length of service, both were kept on—one was moved to another park while the other remained at Pearson Park and was assisted by a much younger man. The delicacy with which this situation was handled implies a parks service that was accustomed to humane and considerate treatment of loyal employees where possible.

Where workers had families and no friendly society membership, they were often paid partial or full salaries for a fixed period of time. In Hull, a parks labourer aged 65 was paid a gratuity of one year's wages and half wages for another six months after becoming paralysed (HHC, Parks and Burials Committee minutes, Volume 6, p. 25). Similarly, the wife and family of a gardener at Headon Road cemetery in Hull were awarded one year's wages after his death at the age of 29. The parks superintendent remarked that he had been in the service of the committee 'since a boy' (HHC, Parks and Burials Committee minutes, Volume 6, p. 55).

Similarly, H. B. Bland, a Liverpool park-keeper aged 38, was sent to Buxton for six weeks in 1893 to recover from a bad attack of rheumatism, which left him 'scarcely able to perform his duties' (LA, Parks and Gardens Committee minute books, Volume 19, p. 103). He had worked in the department for ten years and, on his return, was employed as an occasional man at a reduced salary. He had been paid at half wages for the duration of his time in Buxton.

The most common illnesses suffered by parks workers tended to reflect the nature of their jobs—hard, physical labour, undertaken mostly out of doors. These included bronchitis, gastritis, asthma, rheumatism, furunculosis (skin boils) and influenza. Some parks employees also struggled with alcoholism. In Liverpool, a Sefton Park park-keeper, William Disley, was demoted to labourer due to continual episodes of drunk and disorderly behaviour in 1913 and 1914 (LA, Parks and Gardens Committee minute books, Volume 25, p. 78). By 1916, Disley had redeemed himself sufficiently to warrant his reinstatement as a park-keeper at Rupert Street Recreation Ground (LA, Parks and Gardens Committee minute books, Volume 25, p. 230).

Many employees remained effective workers beyond the age of 65— James Swarbrick, a Liverpool park-keeper, requested permission to retire at age 73 in 1904 (LA, Parks and Gardens Committee minute books, Volume 23, p. 306). All sickness of Liverpool parks employees was recorded along with details of length of service and dates of previous absence due to illness (LA, Parks and Gardens Committee minute books, Volume 30, p. 27). As employees aged, they were often relegated to less onerous tasks, such as watching and opening and closing entrance gates. In 1918, Liverpool employed 44 such men with an average age of 70 years (LA, Parks and Gardens Committee minute books, Volume 29, p. 57).

Decisions to terminate employment were not taken lightly. In Manchester in 1936, a parks labourer, T. Howarth, was eventually removed from employment after suffering several spells of sickness (nervousness and neurasthenia [fatigue and depression]). He had been transferred around seven different parks in the city but had frequently had disputes with other parks staff and had been assessed by a doctor as suffering from a 'mental instability' (MA, Parks and Cemeteries Committee minutes, Volume 51, pp. 85–6). In Hull, Riby Nicholson, a labourer in Hull's

West Park, was dismissed as he had gotten into debt and had spent two days in jail. His creditors had begun to arrive at his workplace to pressure him for their money, which prevented him from doing his work (HHC, Parks and Burials Committee minutes, Volume 6, p. 52). He had also been in trouble the previous year for selling tickets illegally for the annual soiree of Hull Corporation Employees Recreation Society (HHC, Parks and Burials Committee minutes, Volume 6, p. 2).

The length of service was also an indication of the attractiveness of the work for the employees. The relative security of employment, combined with the regular and sustained programme of maintenance, development and repair that most city parks needed, meant that these positions were valued and sought after. However, many staff were reluctant to take time off work for illness and some returned to work when they had not quite recovered. In Salford, Wilsher commented that one workman called Philip Fisk had returned to work in 1900 when he was still 'in a very weak state' (SA, Parks Superintendent Report Books, Volume 4, p. 115). His return had undoubtedly been driven by a fear of either a pay cut or no wages at all or by a desire to prove himself fit and capable of work. Workmen who died in service and who had accumulated many years of service often had a lump sum of money settled on their widows and children by the city council as a reward for their loyalty. In most cases, this was a one-off payment, which, while welcome, was probably not sufficient to make up for the loss of the main wage earner.

Some concessions were granted to parks employees, especially those in search of more training. In 1922, Hull City Council granted its gardener employees the use of a hut for meeting in West Park in the city. The purpose of the meetings was to form a guild to encourage the gardening profession (*Hull Daily Mail*, 23 November, 1922, p. 8). Salford staff who were receiving training in horticulture in the 1930s were granted the use of a library of horticulture textbooks paid for by the parks department and made available to staff in the office at Buile Hill Park (SA, Parks Superintendent Report Books, Volume 11, p. 236). Efforts were now being made to offer support to those staff who were committed to pursuing a career in this area. The availability of such opportunities enhanced the standing and attractiveness of these posts.

The importance of a broader, general education for the lowlier parks employees was also emphasised. Apprentices and young men employed in Manchester's parks were permitted to attend day continuation classes for one day per week to study English, maths, geography, horticulture and science in 1938 (MA, Parks and Cemeteries Committee minutes, Volume 52, p. 129). These improved prospects were also extended to administrative staff in the parks departments in some cities—after a period of 12 months' employment, staff in Manchester were permitted to study for a BA in administration at Manchester University (MA, Parks and Cemeteries Committee minutes, Volume 51, p. 69). Many benefits of the job were offered to all staff, such as the annual staff outing for half a day enjoyed by Salford parks employees in 1936. They also benefited from membership of the Salford Parks Social and Athletic Society, intended to encourage social interaction and leisure activities outside of working hours (SA, Salford Parks Subcommittee minutes, Volume 13, p. 267).

The physical nature of their work was also reflected in the regular use of animals, such as horses, in the parks. Salford City Council had the use of a horse for the heavier aspects of work in the parks, but the horse also felt the after-effects of the nature of the hard labour involved, as he had to be retired due to his inability to carry out the work. The regular nature of the vandalism that occurred in the parks ensured that a programme of repair and maintenance had to be carried out all year around. This relied on a contingent of fit and healthy men available and willing to undertake this work in all weathers. The nature of the work meant that, inevitably, younger men found it less arduous, while older men struggled. There was little room for sentiment, however. Any man who could not complete his work was demoted and someone younger took his place. This practice was by no means confined to parks departments but endemic in all works departments of local councils and municipal authorities.

Annual reports were made by the parks superintendent on the professional performance of each member of staff, and levels of pay increases were suggested accordingly. This relied on a supply of accurate and detailed information about each of the men, and the careful assessment of their contribution and eligibility was made. In cases of employees leaving or dying, staff would be reallocated and often promoted from within. A strict hierarchy was used as regards skilled and unskilled roles. Those who were unskilled were generally older (an average age of 70 in Liverpool in 1918) and undertook menial tasks, such as opening and closing gates and watching (LA, Parks and Gardens Committee minutes, Volume 29, pp. 57–58). The Old Age Pensions Act of 1908 provided for a retirement age of 70 years and a non-contributory, means-tested pension of five shillings per week (Fraser, 2009, p. 183).

All of this required the parks superintendent to have a good working knowledge of the men and their abilities and to be able to judge who was worthy of promotion. Increasing numbers of parks in many British cities and towns in the early twentieth century placed additional pressures on the administration of the service, as well preparing for new innovations, such as superannuation schemes and employee training and development. All employees had to be carefully managed and supervised. Certain types of traits were regarded as necessary to successfully undertake such work in the parks. Commenting on the quality of the employees in Liverpool's parks in 1875, the clerk of works strongly urged the Parks and Gardens Committee not to engage either an ex-policeman or an old soldier as 'we have specimens of both . . . and experience shows that they are idle and always scheming how to shirk what little duty they have to do' (LA, Parks and Gardens Committee minutes, Volume 9, p. 372).

In Liverpool's parks, a bell was rung to denote the time that employees were expected to commence work. The park foreman was tasked with taking the names of the men who were present in his park and ready to begin work. Five minutes' grace was allowed and anyone not present after that time was stopped either a quarter or half a day's pay (LA, Parks and Gardens Committee minutes, Volume 12, p. 284). Careful record keeping was a feature of the management of many parks employees. The city of Hull introduced a system of time books for each park in 1896 that recorded the hours worked by every employee each day. The Parks Committee inspected these books at each meeting (HHC, Parks and Burials Committee minutes, Volume 6, p. 34).

Allied to this careful administration and monitoring of staff were the salaries offered for the job. All cities kept track of equivalent salaries in parks departments and frequently compared rates in neighbouring cities for the purposes of ensuring that rates for the job were alike. In 1917, Liverpool undertook a detailed exercise to compare the salary rates and overtime paid to employees in parks in other cities, such as Manchester, Salford and Chester. The minimum wage in Liverpool parks was 25 shillings per week; it was 26 in Manchester and Salford and 24 in Chester (LA, Parks and Gardens Committee minutes, Volume 28, pp. 412–414). Liverpool paid only five pence an hour overtime, while Manchester and Salford both paid six pence (Manchester paid seven pence to gardeners). This was remedied when Liverpool raised its overtime rates to eight pence as a result of this exercise (LA, Parks and Gardens Committee minutes, Volume 28, p. 417).

In the early years of the twentieth century, many British parks departments took advantage of the opportunity to use unemployed men to labour in the parks, digging new lakes and adding other new features to the landscape. This had been facilitated by the introduction of the Unemployed Workmen's Act of 1905. This act established a degree of state responsibility for the unemployed and 'the first in a line of legislation which ultimately produced the British welfare state' (Brown, 1971, p. 629).

The provisions of the Act were not always successful, as the unemployed men did not always possess the skills needed for the work. Wilsher in Salford complained to the Parks Committee in 1921 that the class of unemployed men was 'very unsatisfactory' and 'most unsuitable' for the type of work involved (SA, Parks Superintendent Report Books, Volume 7, p. 220). The following year, he again appealed to the Parks Committee not to use the unemployed men in the parks department as there was 'no work in the department which could be undertaken with advantage by the class of labour proposed to be taken on' (SA, Parks Superintendent Report Books, Volume 8, pp. 3–4). Efforts were also being made to offer some employment to disabled ex-servicemen in cities such as Liverpool. Where two candidates were equally qualified for a job, preference was given to those who were ex-servicemen and, especially, to those who were disabled (LA, Parks and Gardens Committee minutes, Volume 27,

p. 227). Such men were often employed during the summer months to provide light duties, such as watching, picking up waste paper and weeding flower beds and borders (LA, Parks and Gardens Committee minutes, Volume 27, p. 375).

Many posts in Britain's public parks were seasonal in nature. This was especially the case in the summer months, when parks were experiencing maximum usage and additional employees were needed. The Salford parks department made the decision in 1933 to employ games attendants specifically to help to manage the public's sporting demands and to relieve some pressure from the permanent parks staff (SA, Parks Superintendent Report Books, Volume 11, p. 6). Twenty new games attendants began work in May 1933 and were fitted with light twill khaki jackets to mitigate their untidy appearance (SA, Parks Superintendent Report Books, Volume 11, p. 6). This reflects the importance attached not only to the posts but also to a determination that all parks staff should present a professional appearance to the general public. These uniforms made them easy to identify and were an important symbol of their authority. Eighty extra employees were hired in Manchester in 1938 for the summer season along with an additional 13 extra watchmen (MA, Parks and Cemeteries Committee minutes, Volume 52, p. 74).

Salford decided to make a clear distinction between gardening staff and watching staff in 1934. Watching in parks was now such a vital task due to the crime rates that it could no longer be combined with gardening. Watching was separated from gardening, and 21 men from the bottom of the gardening grade were reclassified as watchmen, to work 47 hours a week and to wear a uniform similar to park-keepers (SA, Parks Superintendent Report Books, Volume 11, p. 164). These men were officially graded as watchmen and were encouraged to pass the St John's Ambulance Brigade ambulance test. Parks were not only becoming more popular and well attended but also posing increasing problems of social order and required more careful and dedicated monitoring.

Such was the scale of these problems in some British cities that consideration was given to using policemen to patrol the parks and not parks employees. Liverpool City Council considered this question in 1919 and wrote to the Home Office to ask for advice with regard to the persistent 'disorderly conduct and offences against decency' (LA, Parks and Gardens Committee minutes, Volume 26, p. 127). The answer received stressed that policing parks should be the responsibility of the Parks Committee and that any policing done by the regular police force would have to be paid for by the Parks Committee (LA, Parks and Gardens Committee minutes, Volume 26, p. 127). Clearly, it was more economically efficient for the Parks Committee to use their own employees for patrolling and watching duties and this tactic was employed in most cities and towns. The additional intervention of the police force was resorted to only in extreme or unusual circumstances.

Outside of World War I, the role of women in administering and managing Britain's public parks was minimal. Cranz has remarked that women were valued more for their 'stabilising influence' than as active partners in their running (Cranz, 1980, p. 82). The perception persisted that the solution to behavioural problems in parks was the employment of women as patrollers. In 1923, the East Toxteth Unionist Women's Association wrote directly to Councillor Mabel Fletcher (one of two women members of the Parks and Gardens Committee) to ask for the employment of paid female patrollers in Sefton Park to 'prevent conditions about which numerous complaints have been made by people in the district'. Another similar letter from the Women Police Propaganda Committee made the same request. The committee rejected both as serving no useful purpose (LA, Parks and Gardens Committee minutes, Volume 30, p. 415).

Being employed in a public park was a job that brought with it certain risks, as well as plenty of physical demands. Park-keepers were frequently verbally abused and sometimes physically assaulted by members of the public in the course of carrying out their duties. In 1928, Salford Parks department took the step of appointing three park rangers to undertake watching duties in the larger parks of Buile Hill and Peel. They wore a uniform and were sworn in as special constables with powers of arrest (SA, Parks Superintendent Report Books, Volume 9, p. 116). In part, this was to prevent direct assaults on park-keepers and to allow them to concentrate on the horticultural aspects of their work. It was also intended to signal a new authority to the public who visited the parks. Park-keepers frequently complained that one of the reasons why the public attacked them was that they had no real authority to detain those who misbehaved. It was also a tacit acknowledgement that criminal behaviour in parks was continuing to cause problems.

Two park-keepers who had been physically assaulted in Salford's parks in 1925 had seen both cases against their perpetrators dismissed due to contradictory evidence presented. The parks superintendent reported that the park-keepers were very upset that the cases 'had been so lightly dealt with' (SA, Parks Superintendent Report Books, Volume 8, p. 225). Another incident of assault on a Liverpool park-keeper necessitated some assistance from members of the public when the assailant continually knocked the park-keeper to the ground. The park-keeper later complained that he 'had not been provided with a staff [stick] to defend himself from violence of this kind' (LA, Parks and Gardens Committee minutes, Volume 27, p. 28).

Complaints from members of the public about the attitude of parks employees were reasonably common. In Liverpool in 1890, a complaint was received from a member of the public who had been prevented from leaving Sefton Park by an assistant park-keeper (Charles Hill) who had locked a gate and prevented his exit (LA, Parks and Gardens Committee minutes, Volume 17, p. 172). The parks superintendent investigated the incident and cautioned Hill 'to avoid any likely charge of incivility' in his future dealings with the public. There was a certain sensitivity in dealing with park users, many of whom regarded the spaces as 'theirs'. In 1906 also in Liverpool, a park-keeper was found drunk in Devonshire Place Recreation ground, having locked a number of screaming children in the park. He was subsequently dismissed from his post (LA, Parks and Gardens Committee minutes, Volume 24, p. 398).

Some parks employees were assigned tasks to which they objected. Three park employees at Stanley Park in Liverpool refused to sweep up leaves when told to do so by the superintendent. They were suspended for a week, the superintendent remarking that 'an insubordinate spirit had been shown' (LA, Parks and Gardens Committee minutes, Volume 20, pp. 183–4). Arthur Fisk, caretaker of a small Salford park, gave his notice when he was asked to clean and scrape the park railings before they were painted (SA, Parks Superintendent Report Books, Volume 8, p. 65). Similarly, two park-keepers at Liverpool's Sefton Park refused to clean the closets there in 1875, claiming that it was not part of their duties and that they had 'delicate stomachs' (LA, Parks and Gardens Committee minutes, Volume 9, p. 217). The men were told by the clerk of works that they would be dismissed if they did not do so. An impromptu strike of parks employees took place in Salford in August 1918 when 25 park workmen walked out without any notice. The parks in which they worked had to close their tennis and bowling greens, and the police were used to patrol so that the parks could remain open. They remained on strike for 11 days (SA, Parks Superintendent Report Books, Volume 7, pp. 33–36).

As well as managing their own staff and liaising with the wider municipal authorities, the parks superintendents were expected to deal with public complaints and liaise with trade union officials. The interests of the parks employees were monitored regularly and their terms and conditions regulated, often by local bodies, such as the Lancashire and Cheshire Whitley Council Local Authorities Non-Trading Services. These bodies were responsible for negotiating pay rates and working conditions for manual workers in hospitals and county councils. Any deviations from the norm were queried. In 1915, the district secretary of the Municipal Employees Association wrote to the Preston Parks and Baths Committee to ask why parks employees began work at 6:30 a.m. when those in all other departments began at 7:00 a.m. (LRO, Parks and Baths Committee minutes, Volume 8, p. 318). All such discrepancies had to be justified or altered accordingly.

In Manchester in 1936, a letter of complaint was received from the National Union of General and Municipal Workers about the recent dismissal of a number of temporary staff in parks and cemeteries (MA, Parks and Cemeteries Committee minutes, Volume 50, p. 218). The parks superintendent replied that, while efforts were made to keep the employment of temporary staff as regular as possible, the work was intermittent by its nature and impossible to predict. The department had tried to keep in mind details such as length of service and qualifications when dismissing temporary staff as their lack of formal training meant that they could undertake only certain kinds of work, such as general construction and maintenance (MA, Parks and Cemeteries Committee minutes, Volume 50, p. 219).

General employment in a public park was a sought-after occupation. These were demanding jobs but ones that provided stability and a reasonable expectation of continuity and training. They were hierarchical but a committed and reliable worker could progress and rely on a reasonable standard of living and income.

Conclusion

Any discussion of the staff of the public park at this time is hampered by the dominance of available evidence from park superintendents and directors and a corresponding lack of material emanating from the gardener and the labourer. The accounts of park-keepers are similarly rare. Thus, any attempt to tell the story of the parks employee is limited. Extrapolations can be made from existing sources as to the ages, illnesses and working experiences of these employees, but personal details and their own voices are missing. During the period of this study, most of the cities saw an increase in the numbers and sizes of their parks. This presented a more complex administrative challenge for which most parks superintendents were very suited. The municipalities were responsive to requests for more administrative support and for more staff in the parks themselves. It is striking to note that the bureaucratic class expanded and became more specialised (a key component of urban culture at this time) while the job specifications of the ordinary parks staff remained quite static. More of them needed to be employed with the additions of new parks, but the mainstays of their jobs remained similar.

The main themes to emerge here are the physical burdens and personal dangers experienced by the parks worker and the increasing dominance of the professional parks director/ manager in the municipality. These key roles ensured the smooth functioning of the entire parks department (with the occasional exception in Hull, which was rare) and the emergence of a notable public official who not only had professional expertise in horticultural matters but also was a skilled and able administrator, capable of managing staff and elected officials alike. Dynastic families, such as the Pettigrews, were relatively rare but an indication that these skills were increasingly valued in the twentieth-century city, and many parks administrators became public figures in their own right.

Much of the authority in the public park derived from the park managers and from those whose job it was to patrol and supervise them. This was an area of municipal endeavour that was growing increasingly complex during the period of this study. The acquisition of more parks and recreation grounds meant a need for more staff and more training programmes. By 1934, the city of Liverpool was spending £100,000 per annum on the maintenance of its parks, three quarters of which went to employees' wages. Net revenues at this time were £30,000 from 113 parks, spanning 2,150 acres (O'Mahony, 1934). Many of these employees have remained invisible from previous park histories, which have tended to concentrate instead on the design, layout and provision of amenities. However, the stories of these men and women can shed new light not just on the day-to-day challenges of working in these spaces but also on the broader trends and factors that impacted on the development of civic authority itself, especially during the civic boom of the early twentieth century. This period also witnessed an increasing demand for a diverse and commercialised public leisure landscape—this topic is the focus of the next chapter.

References

Brown, K. (1971). Conflict in Early British Welfare Policy: The Case of the Unemployed Workmen's Bill of 1905. *Journal of Modern History*, 43(4), 615–629.

Chandler, J. (2007). *Explaining Local Government: Local Government in Britain Since 1800*. Manchester: Manchester University Press.

The City Parks (1915, 27 April). *Manchester Courier*, p. 6.

Corporation Clear-up (1930, 20 October). *Hull Daily Mail*, p. 1.

Cranz, G. (1980). Women in Urban Parks. *Signs: A Journal of Women in Culture and Society*, 5(3), 79–95.

Fraser, D. (2009). *The Evolution of the British Welfare State*. Basingstoke: Palgrave Macmillan.

Garrard, J. (1983). *Leadership and Power in Victorian Industrial Towns 1830–1880*. Manchester: Manchester University Press.

Garrard, J. (1995). Urban Elites, 1850–1914: The Rule and Decline of a New Squirearchy? *Albion: A Quarterly Journal Concerned With British Studies*, 27(4), 583–621.

Hanson, C. (1972). Welfare Before the Welfare State. In R. M. Hartwell, G. E. Mingay et al (Eds.). *The Long Debate on Poverty: Eight Essays on Industrialisation and the 'Condition of England'* (pp. 111–139). London: Institute of Economic Affairs.

Hopkins, E. (1979). *A Social History of the English Working Classes, 1815–1945*. London: Edward Arnold.

Hull History Centre, Kingston-Upon-Hull, Minutes of the Parks and Burials Committee, TCM/2/14-44/6.

Hull Parks Staff (1922, 23 November). *Hull Daily Mail*, p. 8.

In Heaton Park: A Great Holiday Throng (1902, 27 June). *Manchester Guardian*, p. 6.

In the Hull Parks (1902, 24 April). *Hull Daily Mail*, p. 4.

Kidd, A. (2002). Civil Society or the State? Recent Approaches to the History of Voluntary Welfare. *The Journal of Historical Sociology*, 15(3), 328–342.

Lambert, D. (2005). *The Park Keeper*. Swindon: English Heritage.

Lancashire Record Office, Preston, Parks and Recreation Grounds Committee Minutes, CBBU/39.

Liverpool Archives, Liverpool, Parks and Gardens Committee Minute books, 352/MIN/PAR/1/.

Manchester Archives and Local Studies, Manchester, Parks and Cemeteries Committee Minute Books, GB127.Council Minutes/Parks and Cemeteries/1-53.

Manchester: A Continued Celebration (1902, 28 June). *Manchester Guardian*, p. 5.

Mr. H. B. Witty (1929, 7 November). *Hull Daily Mail*, p. 11.

O'Mahony, M. (1934). *Official Handbook: The Parks, Gardens and Recreation Grounds of the City of Liverpool*. Liverpool: Liverpool City Council.

Parks (1929, 11 January). *Manchester Evening News*, p. 7.

Parks Regulation Act, 1872, 35 & 36 Vict., c.15.

The Parks For 'the People!' (1881, 2 July). *Punch*, p. 64.

The Parks: Saturday Afternoon in Heaton Park (1903, 25 July). *Manchester City News*, p. 3.

Pettigrew, A. A. (1926–1932). *The Public Parks and Recreation Grounds of Cardiff*. 6 Volumes. Unpublished.

Pettigrew, W. W. (1937). *Municipal Parks: Layout, Management and Administration*. London: The Journal of Park Administration.

A Question of Power (1930, 18 December). *Hull Daily Mail*, p. 1.

Salford Archives, Salford. Parks Superintendent Report Books. L/CS/DR4/1-13.

Salford Officials Salaries (1902, 1 February). *Manchester Evening News*, p. 2.

Waller, P. (1983). *Town, City and Nation: England 1850–1914*. Oxford and New York: Oxford University Press.

West Yorkshire Archive Service, Leeds. Leeds Parks Committee Minutes, LLC47/1/.

Willes, M. (2015). *The Gardens of the British Working Class*. New Haven and London: Yale University Press.

5 Ideal Playgrounds?
Parks and the Development of Popular Recreation

Introduction

Urban parks made an important contribution to the World War I effort and one that is neglected in existing histories. These parks provided sites for the training of troops, buildings and rehabilitative space for those convalescing from injuries and valuable plots of land where food could be produced. Their importance as peaceful havens achieved a new significance during wartime, when they provided a welcome escape for a war-weary public. The interwar years mark a considerable advance in the development of a commercial leisure market. Greater social freedoms for women meant increasing demands for new forms of recreation in public parks, as well as a more sophisticated understanding of the leisure needs of children, especially the poor.

Changing perceptions of public health and a more relaxed attitude to the human body increased the potential appeal of the urban park as a centre of health and fitness. Popular entertainments, such as dancing, concert parties and the broadcasting of football matches, resulted in public parks that appealed to a broader demographic but one that was more discerning and whose tastes were often fickle and difficult to predict. Municipal park policy was often slow to adapt to the needs of the wartime and interwar park user, but the challenges of this period resulted in some of the most fascinating developments in public leisure.

Over the past few decades, the study of the British public park has concentrated on its origins and immediate Victorian history, such as that produced by Hazel Conway (1991). Our understanding of later developments is patchy and largely confined to studies of park design. The lack of attention is significant in so far as it hampers our comprehension of the continuing contribution of public parks to the evolution of the use of public space for the purposes of leisure and how the tensions that emerged from this usage can be employed to assist us in assessing the success of these spaces. In order to address this, this chapter focuses on how and for what purposes the early twentieth-century public park was used, not just for sporting facilities but also for general entertainment. There is a need to examine both similarities and disparities in the ways in which parks evolved as recreational spaces and the role they played in the development of leisure patterns in a variety of British cities.

Rational Recreation and Beyond: Sport and Leisure at the Turn of the Nineteenth Century

Sporting activity had long been a feature of the public park since its inception in the 1840s. Indeed, sporting facilities located in public parks predated the wider municipal provision of leisure amenities—hence their importance as an object of study (Meller, 1976). At this time, such sports were confined to genteel pursuits, such as golf, bowls and tennis. In 1872, speaking at the opening of Roundhay Park in Leeds, the mayor and principal funder of the park, John Barran, referred to the space as 'an ideal playground for the people of this town' (Burt, 2000, p. 27). Much of the impetus for such developments was 'rational recreation', underpinned by ideas about self-improvement and, later, citizenship. Many other Victorian public parks got a similar approbation on their opening, but as the nineteenth century went on, such ideals faded.

The later nineteenth and early twentieth centuries saw a growing trend towards the maximisation of commercial income in parks and the attempt to incorporate sporting activity in parks into the wider municipal public leisure landscape. Open-air sport was often weather-dependent and began to lose out to commercial pleasure grounds, dance and music halls and cinemas. Hannikainen (2016) has noted that some London county councils, such as Bermondsey, tried to compensate for this by installing floodlighting in one of their parks to encourage winter football. However, there is no evidence that any of the authorities in this study contemplated something similar. Sport, which had once been a strength of the British public park, thus found itself increasingly sidelined, just one of a growing number of possible activities in the leisure-dominated city.

The later renaissance of sport in parks was motivated by the new doctrine of health and physical culture, which became especially popular during the 1920s. A new emphasis on the health of women and children (especially the urban poor) meant a realigning of the role of sport in public parks with new ideas and ideals about the body and physical fitness. Open-air swimming baths, running tracks and pitch and putt were introduced into the landscape of the public park during the 1930s, along with more commercially focused plans to lease parts of public parks for open-air boxing, athletics and show-jumping events. This gave rise to a new tension between the uses of public parks for entertainment, as well as preserving their public leisure function. This tension has much to tell us about changing concepts of public space, the demands of a leisure-hungry public and the role of sport more generally in the urban environment.

While elements of a modified rational recreation were still observable in municipal leisure during the early twentieth century, this chapter suggests that it was public demand and the existence of an increasingly sophisticated and competitive marketplace for leisure that were really driving sporting initiatives. By the early twentieth century, the larger public parks seemed to offer ideal spaces for the accommodation of many diverse sports, along with music and refreshments. The capacity of musical entertainment to attract crowds to the parks, especially at weekends, was considerable. In Manchester's Heaton Park in 1909, local businessman William Grimshaw established a series of gramophone concerts, which proved extremely popular. Grimshaw, a regular attender of classical concerts at the Free Trade Hall, used these occasions to play his extensive record collection on a specially adapted gramophone at the park to crowds estimated at 40,000, an idea that was later emulated in parks in other cities (*Manchester Courier*, 1909).

However, urban leisure enthusiasts were beginning to turn away from public parks and to avail themselves of city centre attractions, such as dance halls and cinemas. The lure of the seaside resort and the rural day trip was also significant in drawing people out of the urban environment altogether. This was impacting on the ability of the urban parks to generate income. In Manchester in 1891, Martha Jane Andrew, the tenant of Queen's Park refreshment rooms, wrote to the Parks and Cemeteries Committee to alert them to the fact that she was struggling to make a living from refreshments as too many people were availing themselves of cheap rail excursions (MA, Parks and Cemeteries Committee minutes, Volume 12, p. 10). In consideration of her situation, the committee agreed to reduce the rent she was being charged for her tenancy.

Rational recreation had tended to view physical activity as a kind of safety valve for the working classes. Much of the development of parks and other cultural institutions, such as art galleries and museums, was spearheaded by middle-class philanthropists and civic leaders—the new 'urban aristocracy' (Briggs, 1968, p. 94). The gradual retreat of the middle classes to the more desirable suburbs in the 1880s left the city populations dominated by the working classes (Savage and Miles, 1994). Therefore, the cultural capital of the urban middle classes was somewhat outweighed by their reduced numbers, but their influence over sporting and leisure developments continued.

In the later nineteenth and early twentieth centuries, British parks authorities busied themselves in both developing existing sports facilities and introducing new amenities. By the 1880s, many parks authorities in towns and cities had begun to develop a competitive instinct about the sports facilities they offered. In 1888, Salford's parks superintendent reported that Manchester's decision to reduce the fee charged for bowling in its parks had affected attendances at the bowling greens in Salford's Peel Park (SA, Parks Superintendent Report Book, Volume 3, p. 122). Contiguous parks authorities had frequently taken the opportunity to learn from the experiences of other park managers as the regards the provision of particular sporting facilities and their popularity. In 1910, Manchester's parks superintendent, William Wallace Pettigrew, published a comprehensive report about the provision of sporting facilities in parks in London, Leeds and Birmingham, among others (MA, Parks and Cemeteries Committee minutes, Volume 29, p. 152). At the time, Manchester was considering how to provide space for sports such as football and cricket in its parks. Now, however, there is evidence of a more market-driven and competitive approach to the pricing and location of these amenities.

In part, this was being driven by the attempt to make public parks year-round destinations. Parks had always been most popular during the summer months and, especially, on bank holiday weekends. The winter months and inclement weather during the year were often a deterrent to parks visitors, and managers began to seek new opportunities to entice people to the parks more frequently and for regular purposes. This reflects an awareness of the fact that parks now had to compete with other, commercial forms of leisure in the urban environment. One response to bad weather was to develop indoor leisure activities, such as games rooms, but these were often used for gambling and some had to be closed as a result (MA, Parks and Cemeteries Committee minutes, Volume 42, p. 148).

Another alternative was to make a virtue of cold or inclement weather by providing facilities for ice skating, ice hockey and curling, which proved popular activities in many parks (MA, Parks and Cemeteries Committee minutes, Volume 40, p. 16). The Manchester Ice Palace, opened in 1910, held dances on the boarded-over ice rink during the summer months (Inglis, 2004). As well as facilities for golf, tennis and bowling, other sporting activities were introduced. Cardiff parks permitted baseball to be played from 1905, which proved to be as popular as a spectacle as well as a sport. Many games attracted crowds of up to 6,000, and there were as many as 30 baseball pitches in Cardiff's parks by 1930 (A. Pettigrew, Volume 3, 1931, pp. 41–42). These pitches were used for football during the winter months in order to maximise their usage.

Parks had always been important arenas for spectatorship since their inception. People visited to see and admire the planting schemes or to listen to a band. The introduction of sports into the environment of the park, however, increased the numbers of people who visited parks with specific kinds of sporting spectatorship in mind. While this may have had the impact of attracting more people to the parks, most were not paying to watch and therefore did not have the effect of increasing revenue. Some municipal representatives did not welcome this more passive form of recreation. In Cardiff, Councillor Snook, the chairman of the Parks Committee, vouchsafed the view that 'the need was for inducements to a larger number of the public to take personal exercise and recreation instead of merely looking at others playing' (A. Pettigrew, Volume 4, 1931, p. 103). Such passivity was associated with the idleness so feared by the Victorians and Snook's comment was an indicator that a view still prevailed that rational recreation should be active and purposeful. It may also reflect an awareness that, if more people participated in sport, they would be more likely to have to pay for it, thus increasing revenues.

Some sports, such as football, were not welcomed in many British parks until well into the twentieth century. This may have resulted from a combination of its origins and association with the working classes (Horne, Tomlinson and Whannel, 1999). The perception may also have been a more pragmatic one—the playing of football was frequently adjudged to cause damage to the terrain of the park that was often difficult and expensive to repair. While important as a sport that could be played in winter and therefore result in revenue during a traditionally fallow period in the parks, football was acknowledged to bring its own problems. Demand for football pitches constantly outstripped supply in cities such as Cardiff. The use of parks by the military during World War I and for the production of food in allotments considerably reduced the amount of land available for football (A. Pettigrew, Volume 3, 1931, p. 37).

Public parks had always offered cheap and generally accessible opportunities for sport and leisure. Now they had to compete with other municipal sports facilities, like swimming pools, and with those offered by private entrepreneurs, in an era when even working people had an increased amount of leisure time. Indeed, it is possible to argue that recreation was now beginning to be seen not as the opposite of work, as the Victorians believed, but as a civic duty, which complemented work and enhanced the individual's ability to function effectively in the workplace (Bailey, 1998). Much sporting endeavour was now undertaken, not for reasons of public health or self-improvement but as a hobby that was enjoyable in its own right (Rodrick, 2004).

Public parks offered a location where the city and the citizen could thus develop in tandem—to 'become a self-governing member of a self-governed community' (Dagger, 1981, p. 717). This emphasis on the community and the explicit link between health and well-being marks a transition from the Victorian middle-class moral imperialism of rational recreation to a more general concern with the health of the population as a whole.

The versatility of public parks is demonstrated in their significance as military training sites during World War I. Larger parks, such as Manchester's Heaton Park, could accommodate military encampments and even provide facilities to entertain the soldiers and their families. Relationships with the military authorities were not always cordial. The parks superintendent of Manchester, William Wallace Pettigrew, remarked on the damage inflicted on cricket and football pitches by military trenching operations in 1915, while a military request that same year for free or reduced rate access to the boating lakes in Manchester's Heaton Park was refused (MA, Parks and Cemeteries Committee minutes, Volume 35, p. 164 and p. 223). The basis for the refusal was a comment by Pettigrew that the provision of boating in the park was 'a purely business concern', reflecting the importance of trying to maximise revenue from all parks users (MA, Parks and Cemeteries Committee minutes, Volume 35, p. 164). Not all parks authorities were quite so parsimonious—the Cardiff Parks Committee permitted all soldiers to bathe free of charge in Roath Park Lake (Figure 5.1) and wounded soldiers were allowed to attend matinee performances in the parks for free (A. Pettigrew, Volume 6, 1932, p. 5). The Manchester Regiment's sports day was held at Heaton Park in April 1915 and was accompanied by military bands and a prize giving by local dignitaries. The event included a one-mile race, wrestling, high jumping and a boxing contest and was attended by 20,000 spectators (*Manchester Courier*, 1915). The importance of such sports events for the general public should be not underestimated. Public parks were popular recreation spaces and were capable of attracting many thousands of visitors during fine weather. The hope of the parks authorities was that such crowds would increase the revenue from the parks if they took refreshments as part of their day out, but this is difficult to substantiate.

The provision of such refreshments, usually in purpose-built tearooms, was an important part of the revenue-generating process in many public parks. Commonly, private caterers, under contract to the municipality, provided the refreshments. Parks committees regularly reviewed the

Figure 5.1 Roath Park boating lake, Cardiff.

Source: © The Francis Frith Collection.

prices charged by the caterers, conscious that the public frequently complained about overcharging. In 1919, the city of Liverpool appointed a special subcommittee to investigate the possibility of the Parks and Gardens Committee taking over responsibility for the catering in the city's parks 'to prevent profiteering' (LA, Parks and Gardens Committee minutes, Volume 26, p. 191). The subcommittee studied the catering arrangements in place in other city parks and discovered that, with the exception of Manchester, all cities used private catering firms. Prices in Liverpool were noted to be lower than in Manchester, and the subcommittee resolved to keep catering private (LA, Parks and Gardens Committee minutes, Volume 26, p. 192).

One of the problems facing municipal authorities after World War I was how to entertain their populations and encourage a healthy lifestyle, without committing themselves to the spending of too much public money. Increasingly, local authorities had been providing entertainment in the shape of municipal theatres, but they had also looked to make use of other facilities under their control for public events. This included the use of public baths for swimming galas and the staging of open-air plays in public parks. Such activities brought local authorities into direct competition with private commercial entertainment providers and resulted in a series of costly legal disputes during the 1920s and 1930s, outlined ahead. Public parks became multifunctional spaces as they tried to compete with a growing variety of leisure opportunities and a more sophisticated public as consumers of leisure. However, such developments ran the risk of impeding their usefulness as 'ideal playgrounds' and drawing many British municipalities into conflict with private providers of entertainment.

Legislative Frameworks for Leisure

National legislative frameworks assisted in the newer entertainment-oriented developments that occurred in public parks. The 1925 Public Health Act permitted more leniency in respect of using parks to produce revenue from leisure and entertainment activities. The Act enabled local authorities to rent out portions of public parks to local cricket and football clubs and to charge the public for admission to watch matches (Public Health Act, 1925, section 69). The National Playing Fields Association, also established in 1925, enabled the securing of open space in cities for playing football (Hill, 2002). The use of public parks for the purposes of entertainment continued to be restricted. The care exercised about this emphasises the worry about the perceived use of public space for profit-making activities.

Costume concert parties were not legalised by the Public Health Act of 1925, thus depriving many parks authorities of a reliable and popular form of income and preventing the staging of plays that required costumes and scenery. The concert parties had proved to be so popular that the Provincial Entertainments Proprietors and Managers Association Limited wrote to Manchester's Parks and Cemeteries Committee in 1921 to complain about 'invasion of their particular province and interference with their industry' (MA, Parks and Cemeteries Committee minutes, Volume 40, p. 132). Manchester had begun to hold concert parties in Whitworth Park only the previous month (June 1921), which had attracted nearly 5,000 attendees over four evenings (MA, Parks and Cemeteries Committee minutes, Volume 40, p. 127).

Thus, the use of urban parks for cultural activities continued to be minimal and greater emphasis was placed on parks as playgrounds for sporting pursuits. The 1925 Act did provide for local authorities to own and manage sports grounds for revenue-producing purposes, again emphasising that sports and not entertainment was the more desired use of land that had been bought with public money (A. Pettigrew, Volume 6, 1932, p. 80). William Wallace Pettigrew, Manchester's parks superintendent, was critical of the Act, describing it as 'absolutely useless' and a 'great disappointment' (A. Pettigrew, Volume 6, 1932, p. 80). His main objection was that the Act permitted the renting of public land to private clubs for sports such as cricket and football, which ran the risk of depriving others of access to that space.

During the 1920s and 1930s, a series of legal disputes occurred between the Theatrical Managers' Association (TMA) and various British municipalities. These disputes were largely a response to the legislation described earlier and centred on the desire by many local authorities to contest the restrictions posed by the 1925 Public Health Act. Private commercial entertainment providers, for their part, had become increasingly concerned about the activities of municipal authorities in this area (*Manchester Evening News*, 1926). Cardiff Corporation introduced the Cardiff Corporation Act in 1930, which was designed to allow the municipality to provide concerts, exhibitions and other entertainments in parks, pavilions and recreation grounds (A. Pettigrew, Volume 6, 1932, p. 82). This was a direct challenge to the restrictions of the 1925 Public Health Act and an attempt by the municipality to develop its entertainment-providing capacity. The TMA opposed the bill unsuccessfully, as it continued to do in many British towns and cities, such as Hull, Hastings and Eastbourne (TMA Archives, minute book 1927–1936, p. 169).

The TMA's fear was not just that municipal authorities were encroaching on its entertainments business but also that, as its parliamentary representative Charles Watney commented, 'the municipality which runs variety entertainments today will run theatres tomorrow' (TMA Archives, minute book 1920–1928, p. 219). The growing confidence and assertiveness of the local authorities in respect of providing entertainments were posing a direct threat to the private entertainment entrepreneur. Part of Watney's role was to alert the TMA to the future introduction of new legislation in Parliament that would increase the capacity of municipalities to provide such entertainment themselves. Indeed the vice-president of the TMA had been involved in the drafting of the 1925 Public Health Act and the protection of members' interests enshrined in that bill (TMA Archives, minute book 1920–1928, p. 174).

By 1930, some 90 municipal bills were planned, representing a substantial threat to the TMA members. Some municipalities, such as Southport, already had in place acts that allowed them to stage their own entertainments and that predated the restrictions of the 1925 Public Health Act (TMA Archives, minute book 1927–1936, p. 203). While many bills were withdrawn or amended following TMA opposition (including those proposed by Bradford Corporation in 1927 and London County Council in 1935), local authorities were embarking on a determined pathway to finding new ways of producing revenue from their parks and open spaces (TMA Archives, minute book 1927–1936, p. 344).

Some national legislation impacted adversely on the provision of entertainment in public parks. The Sunday Entertainments Act of 1932 enabled cinemas to open on Sundays and greatly assisted in the boom in the popularity of cinema-going by the late 1930s. The decline in the popularity of music halls fuelled the growth of cinemas in British cities— by 1939, Manchester had 129 cinemas, the largest number of any city outside London (Kidd, 2002). Another significant legislative development for parks and recreation was the 1937 Physical Training and Recreation Act. This act was motivated in part by concern about high rates of unemployment among the young and by Britain's increasingly poor record in international sporting competitions (Henry, 2001). The large sum of £2 million pounds was made available under the Act for local authorities to provide facilities for training and recreation, such as swimming baths, playing fields and camping sites, and to train teachers and leaders to organise and supervise such leisure pursuits (Physical Training and Recreation Act, Ch. 46, Section 3).

Public parks were thus redefining their roles to include the provision of entertainment in its broadest sense. The idea of parks as playgrounds was being interpreted more liberally, and municipalities were seeking greater freedom through the introduction of local legislation, which was designed to avoid the restrictions of some national laws. The character of the Parks Committee was changing also. The early decades of the twentieth century saw the election of female members to city councils around Britain. Manchester's Margaret Ashton was first elected to the city council in 1908 as an Independent member for Withington, joining the Parks and Cemeteries Committee in 1910 (MA, Parks and Cemeteries Committee minutes, Volume 29, p. 238). The first woman to be a member of Preston's Parks and Baths Committee was Mrs Councillor Pimblett, who was elected in 1920 and remained a member until 1932 and her promotion to alderman (LRO, Parks and Baths Committee minutes, Volumes 9 and 10, p. 32 and p. 27). Manchester Parks and Cemeteries had its female member from 1932, Mrs Nellie Westcott, a Conservative councillor who was also involved with a number of industrial schools (Hunt, 2012). In 1934, she was joined by two additional women, Councillors Hannah Mitchell and Ellen Griffiths (MA, Parks and Cemeteries Committee minutes, Volume 49, p. 166).

Women's ability to play an active role on city councils was constrained by assumptions about their interests. Margaret Ashton was refused the opportunity to join the Watch Committee of Manchester City Council in 1913. This led the *Manchester Courier* to remark that women were relegated to committees that deal 'with women's affairs. The public health, the parks, the education, are their chief interests' (*Manchester Courier*, 1913). While Jane Bedford has argued that local government 'offered a multiplicity of routes through which women could advance their rights', they were often, in practice, limited by stereotypical ideas about their concerns (Bedford, 1998, p. 9). Hollis's (1987) work has suggested that women found it more difficult to get elected onto city councils unless they already had a pre-existing reputation and experience of council work. Until 1914, they had to be ratepaying electors in order to stand as candidates. Council seats could be valuable stepping stones to parliamentary seats so they were often heavily contested. Once elected, the opportunities for innovation were limited due to the nature of municipal committee business, thereby confining women to 'softer', more female concerns, such as education and public health. Manchester City Council member and suffragette Hannah Mitchell commented on her experience that she did not believe that female councillors had any 'superior administrative capacity' but that their priorities better reflected the concerns of female citizens and that their impact on local legislative agendas was therefore significant (1928, p. 2).

'Ideal Playgrounds'? Recreation Versus Entertainment in the 1920s and 1930s

From the early twentieth century onwards, the impetus of the public parks movement changed from the Victorian idea of rational recreation for the working classes to an emphasis on active citizenship and the importance of all parks users undertaking responsibility for their upkeep. While this made public parks potentially more democratic and inclusive spaces, it also raised questions about the future developments of these parks in the context of the wider city. The emergence and popularity of private commercial forms of entertainment, such as music halls and cinemas, meant increasing pressure on public parks to compete as part of a general regimen of public health and leisure activities. In part, many of the decisions about how to develop public parks were a reflection of a broadening definition of public health, away from specific matters such as sanitation and slum removal and towards issues such as recreation and physical fitness. This manifested itself in organisations like the Manchester Physical Health Culture Society and Leeds's Everywoman's Health Movement, devoted to promoting outdoor sports and physical development.

In 1936, Councillor Tebb, the chair of Manchester's Parks and Cemeteries Committee, condemned plans to lease the city's public parkland for entertainments such as open-air boxing, athletic events and fireworks displays. Tebb remarked that it was the duty of the Parks and Cemeteries Committee 'to find places for recreation and not to be entertainers' (*Manchester Evening Chronicle*, 1936). This comment illustrates many of the dilemmas that faced municipal governments in respect of the provision of leisure. They were attempting to develop such valuable but expensive resources as public parks, while catering to rapidly changing tastes and a very competitive environment. Successful attempts had been made to find places for recreation in frequently neglected areas of British cities, but providing entertainment as well as leisure was proving challenging.

Tebb's comment, in setting recreation against its supposed opposite, entertainment, illustrates the perceived role of public parks as spaces of physical activity as opposed to the more passive entertainment. It also reflects a commonly held suspicion of the more commercial world of entertainment in contrast to the purer and healthier recreation. This indicates that the Victorian attitude to 'rational recreation' remained a relevant touchstone for some municipal representatives.

Many entertainments allowed in public parks mimicked those of the commercial pleasure ground. Such private initiatives, such as that at Manchester's Belle Vue and Crystal Palace in London, had offered a variety of stalls, fairground rides and other attractions since the Victorian period (Pussard, 2004). By the later 1920s, many cities had a flourishing supply of cinemas and music halls so that the private sector provision of popular entertainment was a vigorous competitor to what the municipality could provide. The openings of a speedway (dirt track racing) and a greyhound track in Manchester in 1926 both proved very popular and increased the competition for a leisure-hungry population. Indeed, greyhound racing proved so popular that it attracted 333,375 people in its first year (Jones, 1990).

Other cities show similar patterns in respect of the development of a variety of entertainment facilities. Large super-cinemas and leisure complexes were a feature of many cities during the late 1920s and 1930s. These could hold several thousand people at a time, an example being Liverpool's The Paramount on London Road, opened in 1934 with a capacity of 3,000 (Ackroyd, 2002). Cinemas such as this could also accommodate band concerts and stage shows, which increased their earning capability. The Rialto in Toxteth included a ballroom, café, billiard hall and shops (Ackroyd, 2002). These newer leisure activities were designed for a more structured lifestyle and based on specific times and venues (Stevenson,

1984). This aspect was important in establishing public parks as leisure spaces in their own right. Manchester's parks superintendent, William Wallace Pettigrew, continually emphasised the importance of holding events such as costume concert parties on the same night of the week at the same time to ensure a consistency of attendance (MA, Parks and Cemeteries Committee minutes, Volume 40, p. 127). This was also a recognition of the fact that people were inserting leisure time carefully into their evenings and weekends and were planning recreational activities accordingly.

New attractions like cinemas and dance halls also provided competition not just for the public parks but also for the private commercial pleasure grounds, such as that at Belle Vue, in Manchester (Pussard, 2004). In 1926, permission was granted to the BBC to broadcast the sounds of the English Cup tie final between Manchester City and Bolton Wanderers in Platt Fields Park (MA, Parks and Cemeteries Committee minutes, Volume 43, p. 160). The sounds of the crowd and the commentary from Wembley stadium were broadcast to the park visitors through a series of loudspeakers. Changing fashions in the sporting taste of the public were also being recognised. The grass tennis courts at Platt Fields Park were replaced by a pitch and putt golf course in 1936 (*Manchester Guardian*, 1936).

Revenue generation from leisure facilities in public parks continued to be inconsistent. Some sports were remarkably popular and enduring in their appeal, largely tennis and bowling. These tended to rely for their income on the hire of the courts and greens by local clubs. The bulk of the modest revenues accruing from public parks during the 1920s and 1930s came from sporting activities. Income from other, newer forms of leisure, such as dancing, was meagre. Dancing was a respected social skill and one which was introduced to public parks in some British cities after World War I (Davies, 1992). Open-air dancing at Salford's Ordsall Park brought in £15 over a week in June 1921, while dancing ceased in Manchester's Alexandra and Platt Fields Parks in 1922, having raised only similar amounts (SA, Parks Superintendent Report Books, Volume 7, p. 252; MA, Parks and Cemeteries Committee minutes, Volume 41, p. 77).

There was a general acknowledgement in parks committees that sport would continue to provide most of whatever small incomes there were and that expenditure would generally outweigh income considerably. This is a manifestation of the commitment to invest in public parks as an important component of urban life and as much-needed recreational spaces. In general, expenditure ran at five times the income rate during the late 1920s. In Liverpool, average annual total expenditure ran at around £120,000, while income was rarely above £25,000. Manchester displayed a similar pattern and most of the income in both cities derived from sporting facilities (LA, Parks and Gardens Committee minutes, Volume 32, p. 307 and p. 533; MA, Parks and Cemeteries Committee minutes, Volume 46, p. 37).

While these arrangements provided a fairly steady flow of money, they could also be a source of tension between the parks committees and the clubs themselves (Holt, 1992). These tensions often took the form of what was often perceived by the parks committees as frequent demands for more and better facilities—a constant need to replace and maintain the playing surfaces and to add locker rooms, changing facilities and toilets and the provision of separate facilities for women and men. A new women's pavilion was built at Birchfields Park, Manchester, for female bowlers in 1921 (MA, Parks and Cemeteries Committee minutes, Volume 40, p. 65). Cardiff returfed most of its bowling greens after World War I, in response to requests from players (A. Pettigrew, Volume 4, 1931, pp. 54–56).

Some sports experienced waves of waning popularity, forcing parks committees to consider replacing them with other sports or accepting declining revenues in the hope of a future improvement. Salford's parks superintendent noted a decline in the use of bowling greens and tennis courts in 1926, largely due to an increase in the private provision of both of these amenities (SA, Parks Superintendent Report Books, Volume 8, p. 225). Hull's parks superintendent explained to the city's Parks and Burials Committee in 1937 that interest was waning in bands in the parks. 'Nowadays, people just walk by them', he commented (*Hull Daily Mail*, 1937). It is likely that the private facilities were more numerous and better equipped, thus explaining why they were more attractive. Requests were often received to vary the cost of the hire of the spaces at times when the enthusiasm for particular sports declined. Sefton Park Bowling Club asked for and received a reduction in the rent it paid due to falling memberships in 1932 (LA, Parks and Gardens Committee minutes, Volume 33, p. 187).

During this period, many municipal public parks were offering a mixture of sports and more general entertainments, such as circuses and fun fairs. These temporary activities contributed to the revenue-generating aims of the parks committees and, because they lasted for only a limited period, did not necessarily disrupt the day-to-day use of the parks. Fairground entrepreneur J. J. Butterworth regularly staged his circus at Sheil and Stanley Parks in Liverpool—this kind of entertainment was commonly confined to certain times of the year (summer months) or to celebrate special occasions (a coronation). Three fun fairs were held in Liverpool parks in 1937 to mark the coronation of George VI, the showmen paying a fee to the Parks and Gardens Committee of between £200 and £2,000, depending on the size and duration of their shows (LA, Parks and Gardens Committee minutes, Volume 31, p. 389).

These entertainments were not without problems, however. A deposit had to be introduced in Liverpool in 1935 for those wishing to stage a show in a public park. This money covered the provision of sanitary services and the removal and disposal of refuse after the show. The deposit

system had been introduced after difficulties had been experienced in set-
tling the accounts with the fun fair providers (LA, Parks and Gardens
Committee minutes, Volume 31, p. 10). On occasion, some of the acts
in these shows caused concern—two lions were used in a Wall of Death
performance at a Butterworth circus in Liverpool in 1937, which caused
alarm in the Parks and Gardens Committee about the safety of the animals (LA, Parks and Gardens Committee minutes, Volume 31, p. 489).
There was also periodic concern about the prices charged to the public by
these private operators. Liverpool permitted fun fairs to be held in public
parks only from 1924 and subject to the protection of the public from
being overcharged (LA, Parks and Gardens Committee minutes, Volume
27, p. 528).

Some councillors raised objections to the type of person whom these
fun fairs were deemed to attract. In Liverpool, concerns were raised at
the Parks and Gardens Committee about 'the assembly of an undesirable
class until a very late hour at night' (LA, Parks and Gardens Committee
minutes, Volume 32, p. 282). Such comments serve as a reminder that in
the eyes of many of those charged with the management of public parks,
not all park users were the same and that popular entertainments such as
fun fairs continued to be associated with those who could not be relied
on to behave according to the prevailing social norms. These anxieties
stem from the vestiges of rational recreation and the desire to exert a
form of social control and to police the activities of the working classes
(Bailey, 1978).

Perhaps as an antidote to these fears, cultural events were increasingly
staged in public parks, with Liverpool's Sefton Park hosting the national
Eisteddfod in 1929. This was as much a reflection of the expatriate Welsh
community in Liverpool as it was the existence of a facility such as Sefton Park, which could accommodate such a popular event. Such large-
scale events were easily accommodated in these parks without disrupting
other users. The annual International Sheepdog Trials were held at Stanley Park, Blackpool, at London's Hyde Park and at King's Park, Stirling,
during the 1930s (International Sheepdog Society, 2018).

Some parks users found that their leisure needs were being specifically addressed for the first time. This was particularly true in the case
of children. Parks authorities worked with children's charities to address
their needs directly. In 1922, provision was made for 30,000 children to
visit Manchester's Heaton Park during the summer in conjunction with
Pearson's Fresh Air Fund (MA, Parks and Cemeteries Committee minutes, Volume 41, p. 93). Cyril Arthur Pearson, founder and owner of the
Daily Express newspaper, established this organisation in 1896 (Pearson,
1983). It aimed to take disadvantaged children from cities into the countryside, but this agenda had evolved into the use of urban parks as a rural
substitute. Heaton Park's White Heather Camp provided holidays for
poor children in the parks during the 1920s and 1930s. Pony rides and

sand gardens were designed to facilitate opportunities for young children to enjoy the attraction of the parks in their own right.

Open-air swimming baths were introduced in many public parks during the 1920s. These proved to be extremely popular with all parks users but were not without their dangers. A 12-year-old boy drowned in Hull's East Park in 1921 (HHC, Parks and Burials Committee minutes, TCM/2/34/6, p. 67). The inquest found a need for more supervision of the bathing facilities and that the attendant was 'over-taxed in his duties' (HHC, Parks and Burials Committee minutes, TCM/2/34/6, p. 68). The introduction of mixed bathing to Manchester's parks in 1929 added to the popularity of the baths and to the problems of crowd control (MA, Parks and Cemeteries Committee minutes, Volume 46, p. 47). In July 1930, Pettigrew reported 'a good deal of horse-play by men without lady companions' and suggested the establishment of men-only bathing times (MA, Parks and Cemeteries Committee minutes, Volume 46, p. 191). Despite the behaviour control problems, the open-air bathing was a success, not just for the swimmers but also among the wider public of parks visitors, many of whom gathered near the baths to watch the swimmers (MA, Parks and Cemeteries Committee minutes, Volume 46, p. 191). In 1933, Roundhay Park in Leeds opened a diving stage at Waterloo Lake to enable the parks authorities there to hold swimming and diving competitions (WYAS, Leeds Parks Committee minutes, LLC/47/1/4, p. 239).

The addition of sunbathing facilities by the side of the open-air baths in 1930 was instigated by a request from the Sunlight League, who believed in the health-giving properties of sunshine (MA, Parks and Cemeteries Committee minutes, Volume 46, p. 61). The chairman and deputy chairman of the Parks and Cemeteries Committee, along with Pettigrew, visited Platt Fields Park to investigate the opportunities for sunbathing and allowed the addition of patches of sand by the side of the baths for the bathers to lie on (MA, Parks and Cemeteries Committee minutes, Volume 46, p. 88). A separate enclosure for mothers to sunbathe with their children was also established at Platt Fields Park, presumably to allow some privacy (MA, Parks and Cemeteries Committee minutes, Volume 46, p. 222).

Paddling pools were constructed for children to enjoy; Salford's Ordsall Park opened one in 1928 (SA, L/CS/DR4/9, p. 145) and another at Liverpool Street Recreation Ground in 1931 (SA, L/CS/DR4/10, p. 96). The latter was crowded on its August bank holiday opening, and the parks superintendent reported that 'literally hundreds of children swarmed into the pool as soon as the gates were opened' (SA, L/CS/DR4/10, p. 96). In Liverpool, the Parks and Gardens Committee agreed to share the costs of swimming and paddling pool construction at Stanley Park with the Baths Committee. Their joint report noted that 'Liverpool is not fortunate enough to possess natural bathing facilities on a seafront and it is hoped that the deficiency will be remedied by the proposal' (LA, Parks and Gardens Committee minutes, Volume 30, p. 143). No doubt the hope was also that the existence of such facilities might tempt visitors away from the seaside and into the parks.

Despite these new attractions, many of the sports offered in the original public parks of the 1840s continued to be popular. Some of these sports attracted as many spectators as they did participants. The parks superintendent in Cardiff reported that some eager attendees at baseball games were found to have cut down the ropes that designated the playing areas in order to get closer to the event. They frequently caused injury to the players by leaving their bicycles nearby, which were then tripped over by players running for long balls (A. Pettigrew, Volume 3, 1931, p. 41). Baseball retained its appeal for the sporting public until World War II. Liverpool entrepreneur John Moores helped to establish the National Baseball Association in 1933, and 6,000 spectators attended a baseball match held at Wavertree Park in the city in 1938 (Physick, 2007).

Other sporting events accommodated in public parks were rather more dangerous and consequently appealing to spectators. In 1935, Hull's West Park staged motorcycle racing, in aid of the British Legion (*Hull Daily Mail*, 1935). Motorcycle ownership had increased dramatically during the 1930s and appealed to young men who enjoyed the technical aspects of working on and stripping down the cycles (Holt, 1992, p. 199). Many of these events involved well-known riders and offered a high-octane visual experience for spectators, such as a 'massed start', in which riders lined up on one side of the road and dashed over to their machines when the starting flag fell. Crashes were common on the rough roads of the park—motorcycles ran into embankments and skidded around corners (*Hull Daily Mail*, 1932). Motorcycling clubs were responsible for third-party insurance for the events and for repairing the parkland afterwards (*Western Morning News*, 1939).

One factor that impacted consistently on the development of sporting and leisure facilities in public parks was the preservation of the idea of the sanctity of Sunday. Garrard (1983) has discussed how a proposal to found a public park in memory of Robert Peel in Bolton in 1850 was delayed by six years due to the hostility of the local clergy to Sunday opening. By the early twentieth century, most parks were open on Sundays, refreshments were available and musical entertainments were offered (A. Pettigrew, Volume 6, 1932, p. 61). As the twentieth century advanced, there was recognition that the musical tastes of the park-visiting public were becoming more varied. While military and brass band music remained a feature of parks entertainments, they were complemented by other, more orchestral types of music. This was partly as a result of many parks authorities now employing professional music advisors who were charged with varying the diet of music on offer. The programmes for the 1922 summer season in the parks, by Manchester's musical advisor, W. A. Wilks, included both Pickerill's Orchestra and the Crossley Motor Employees Orchestra as well as brass bands (MA, Parks and Cemeteries Committee minutes, Volume 41, p. 105).

However, the provision of sports on Sundays continued to be outlawed in most British cities. There were periodic attempts to overturn this sentiment, most unsuccessful. The Preston Local Committee of the British Social Hygiene Council communicated a resolution to the Parks and Baths Committee to open the public parks for games on Sundays after 3:00 p.m. (LRO, Parks and Baths Committee minutes, Volume 9, p. 252). The request was refused. The National Secular Society (established in 1866) lobbied Liverpool's Parks and Gardens Committee to open parks on Sundays to no avail (LA, Parks and Gardens Committee minutes, Volume 29, p. 526).

Sunday was preserved as the day when the 'parks as walks' aspect was emphasised. Keeping Sunday free from sporting activity was also central to the Victorian idea of rational recreation and closely linked to the allied temperance movement (Bailey, 1978). The fear was that the availability of sports in parks would damage attendances at churches and Sunday schools. Parks authorities also worried about the cost of providing additional services and facilities on another day of the week (A. Pettigrew, Volume 6, 1932, p. 62).

There was some variation in the toleration of some sporting activities on Sundays—Leeds, London and Birmingham allowed boating on Sundays in 1921 but Cardiff did not. The Cardiff parks superintendent, Andrew Pettigrew (the brother of William), recorded that the other municipal authorities attested that Sunday was the day of the week that represented the maximum income for boating (A. Pettigrew, Volume 6, 1932, p. 62). While public opinion seemed to be broadly in favour of the provision of sporting facilities in parks on Sundays, parks committees all over the country struggled to make a positive case. In some cities, like Liverpool, there was no consensus between the Parks and Gardens Committee and the wider city council on the issue—votes in favour of limited sporting activities in some parks on Sundays were invariably overturned at city council, making any progress difficult (LA, Parks and Gardens Committee minutes, Volume 33, pp. 227, 234, 399, 406 and 414). In 1929, the Parks and Gardens Committee in Liverpool calculated the expense of opening on Sundays as £3,306 per annum in additional wages, while the estimated income was only £2,500, mostly from golf (LA, Parks and Gardens Committee minutes, Volume 29, p. 498). In Hull, a lengthy and controversial debate resulted in a vote to open the municipal golf courses on Sundays in 1930 (*Hull Daily Mail*, 1930).

It is possible to ascribe a variety of motives for the participation in sport and other forms of leisure activities during the time period of this study. These include pleasure, enjoyment, thrills and excitement, feeling part of a community and expressing an identity, the desire to compete and feel physically fit and healthy, to exercise skill and judgement, to become well known and earn money, to avail oneself of facilities and seek new friendships (Tranter, 1998). All of these motives (and more)

expressed themselves in the seeking out of leisure opportunities according to an individual's interests and financial circumstances, in both the
public and private sectors, in the city and at the seaside.

By the mid-1920s, the availability of train and steamboat excursions on
Sundays proved popular with the leisure-conscious public (A. Pettigrew,
Volume 6, 1932, p. 64). These excursions were also linked to post-war
developments in seaside resorts, such as Torquay and Blackpool (Walton,
1983). The ability to escape the confines of the city on a Sunday was
enthusiastically embraced by those who could afford to do so. Indeed,
many of the more popular resorts abandoned the Sunday observances by
the first decade of the twentieth century and offered concerts and other
entertainments, emphasising their difference from the provision of such
pastimes inland (Walton, 1983).

The introduction of some popular entertainments, such as the staging
of plays and musical concert parties, served to dilute the appeal of the
public park as a sporting space and fragmented the visitor experience.
The opening of speedways, racetracks, greyhound and boxing stadia and
other popular spectator sport provision in the cities examined here all
contributed to drawing the general public away from suburban parks
into city centres, which rapidly became the focus for most entertainment
events. Public parks had attempted to reposition themselves as providers of more general entertainment but, in so doing, had broadened their
remit beyond the original forms of 'rational recreation'.

Post-war, there had emerged a new consensus about emphasis on
health and fitness and on body image that was to have consequences for
the urban park. The body began to be perceived as capable of improvement and physical perfection. A healthy and fit body was a hallmark of
a good citizen and central to one's civic duty (Zweiniger-Bargielowska,
2010). Thus, the definition of citizenship became restricted to an association with physical health. Commenting on the popularity of the open-air
baths in parks, the *Manchester Guardian* observed that the water at the
baths at St George's recreation ground in Hulme was 'swarming with
vigorous young bodies which, but for its existence, would probably have
been lounging about the dreary adjoining narrow streets' (*Manchester
Guardian*, 1926). Moreover, a healthy body made for a happy disposi-
tion in the individual. This was reinforced by the 1937 Physical Training and Recreation Act, which enabled the creation of physical training
centres in cities. Many municipal authorities, such as Manchester City
Council, undertook to build physical training centres in the city's parks
that included gyms, drill halls and physical culture rooms. Grants were
available from the Treasury to fund these initiatives.

While the desire for physical fitness was an extension of the Boer War–
era anxiety about the physical condition of the working classes, it was
also indicative of a new movement that emphasised the importance of
physical exercise for all. Initially, such assumptions were predicated on

the links between parks and what the *Manchester Evening Chronicle* described as 'healthy and manly development' (1902). However, associations such as the Women's League of Health and Beauty (WLHB) and the Everywoman's Health Movement were formed and quickly became popular, especially with working women. The Manchester branch had a membership of 3,050 in 1936 and incorporated fitness classes with other pastimes, such as dancing (*Manchester Guardian*, 1936). Such organisations prioritised particular bodily ideals and shapes to aspire to and 'pointed towards a widespread desire among women to emulate the modern body ideal' (Zweiniger-Bargielowska, 2010, p. 240). Mary Bagot Stack, founder of the WLHB, and her daughter Prunella emphasised the importance of fitness for 'racial health' (Zweiniger-Bargielowska, 2010, p. 240). This idea was echoed by other women—in 1933, Sali Lobel of the Everywoman's Health Movement wrote to the Manchester Parks and Cemeteries Committee to ask for permission to give performances in the park similar to the English Folk Dance Society for the purpose of encouraging 'racial health in a radical form' (MA, Parks and Cemeteries Committee minutes, Volume 49, p. 15). Her request was refused and turned down again the following year (MA, Parks and Cemeteries Committee minutes, Volume 49, p. 166).

Physical exertion of the body was becoming an important health indicator and also a significant part of consuming leisure opportunities. Group exercising allowed individuals to promote themselves as healthy, attractive and publicly on display. This view of citizenship was one that emphasised the importance of place and locality. Citizenship was expressed not just at the imperial or national level but also within the confines of a particular town or city.

A comment from the *Manchester City News* expresses neatly the distinction between the Victorian and early twentieth-century attitudes to the entertainment of the public: 'the man who caters for the amusement of the public has to consider the taste of the many-headed, not that of the refined and educated few' (*Manchester City News*, 1908). The discrepancy between the popular and the 'mass' and the rarefied elite mirrors the movement from the early conception of the Victorian public park as a place for genteel contemplation to a complex and contested space.

New influences such as city and urban planning also began to have an impact on how parks were perceived in relation to the wider cityscape. There was a movement in the 1930s to remove park railings and gates and to thereby integrate the landscape of the urban park more firmly into that of the city. Instead of representing a space in opposition to the urban landscape as the Victorian park had done, the early twentieth-century park sought to become incorporated into the cityscape.

In part, this was an extension of the influence of the garden city movement but one that sought to open up parkland vistas and to reflect and accommodate the new urban characteristic of commuting (Conway,

2000). Many cities built parkways—arterial roads that linked city and commercial centres with suburbs and residential areas. Manchester's Princess Parkway opened in 1932. Planned and designed by Barry Parker, with the involvement of William Pettigrew, this road to the garden suburb of Wythenshawe further facilitated the integration of city and parkland, as it bypassed the 60-acre Alexandra Park (*Journal of the Manchester Parks and Cemeteries Staff Society*, 1931–1932).

The intention of the parkway was to create vistas leading off from the road in different directions and to use such roads to awaken the interest of the public in nature. The road was designed to have land and footpaths on both sides, which were planted with trees and shrubs in an echo of the nearby parkland. Thus, the changing season could be observed and appreciated by those undertaking car journeys. At the opening ceremony of the parkway, the minister for transport, P. J. Pybus, suggested that such roads were 'a mere canvas on which the citizens are to paint their own gardens', an interesting comment that again reflects the hope that citizens could and would make the urban landscape their own (*Manchester Evening News*, 1932). Such a zeal for engaging the citizens with their environment was clearly fanciful to a degree but was nevertheless a reflection of the fact that municipal government was attempting to be more proactive and less introspective.

The changing nature of local government politics in the early decades of the twentieth century should also be considered. Fraser (1979) has acknowledged the 1840s as the zenith of municipal improvement activities in many towns and cities, such as Salford, Leeds, Liverpool and Manchester. This also coincided with the development of much public parks provision. Such innovations often necessitated some philosophical consideration of the role and the limits of the local authority, and it was not uncommon to have divisions within and across party lines on the same issue (Fraser, 1979). By the 1930s, the remit of local authorities had expanded considerably to include many functions that had previously been the responsibility of other institutions, such as education and boards of guardians. This rendered much municipal activity part of a complex and bureaucratic machine whose many moving parts were more and more difficult to steer.

Significant changes in patterns of leisure were beginning to emerge during the interwar years—the recreational needs of women (initially discouraged during the Victorian period) and those of children, particularly working-class children, were now being specifically addressed. These were not met in any unified or consistent fashion, but their acknowledgement alone indicates that important changes were underway in how leisure was regarded and in the role it played in the formation of young lives in an urban environment. There remains much to learn about how the leisure interests of both of these groups developed during these years.

Conclusion

From the available evidence, British public parks were far from 'ideal play-grounds' in the early decades of the twentieth century. Having extended their activities beyond the provision of spaces for rational recreation, they struggled to provide sufficiently diverse facilities for an increasingly demanding public. While they continued to be popular recreational spaces, especially at weekends and on bank holidays, they were now just one option among many for the urban seeker of entertainment. Private sector leisure providers could react more quickly to changes in public tastes, while parks committees relied on the proactivity of their parks superintendents and on their ability to convince their committees to act quickly and to invest in the maintenance of existing facilities and the provision of new ones. All initiatives also had to be approved by the wider city councils, who were often acutely conscious of what levels of further expenditure would be deemed tolerable by the ratepayers (Garrard, 1983).

With increasing amounts of free time available for leisure, city dwellers took full advantage of the municipal park but those who benefited most had both the recreational skills and the leisure time to access the spaces. The working classes remained on the periphery and the needs of particular groups of users, such as women, were yet to be fully met until well into the twentieth century. The restrictive atmosphere of the Victorian park gradually eased as responsibility for moral and physical rectitude passed from the park-keeper to the individual visitor. The effect of this was a transfer of emphasis from the passive strollers (whose needs were still accommodated) to the active users, whose various recreational needs could be served simultaneously. The spatial segregation of the larger early twentieth-century urban park emphasised the accommodation of many kinds of leisure and non-leisure activities simultaneously. A new type of diverse cityscape was now capable of serving a new kind of consumer—one whose demands for public leisure facilities were only beginning. This placed more pressure on the public park to sustain itself as an environment in its own right, which forms the subject of the following chapter.

Acknowledgements

Parts of this chapter are derived from an article published in the *International Journal of Regional and Local History* (November 2013) © Taylor and Francis (available online: https://doi.org/10.1179/2051453013Z.0000000009).

References

Ackroyd, H. (2002). *Picture Palaces of Liverpool*. Liverpool: Bluecoat Press.
Bailey, P. (1978). *Leisure and Class in Victorian England*. London: Routledge.

Bailey, P. (1998). *Popular Culture and Performance in the Victorian City*. Cambridge: Cambridge University Press.

Bedford, J. (1998). Margaret Ashton: Manchester's 'First Lady'. *Manchester Region History Review*, 12, 3–17.

Briggs, A. (1968). *Victorian Cities*. London: Penguin.

Burt, S. (2000). *An Illustrated History of Roundhay Park*. Leeds: S. Burt.

Caruso Concert in Heaton Park (1909, 24 September). *Manchester Courier*, p. 3.

The City Parks (1915, 22 April). *Manchester Courier*, p. 6.

Conway, H. (1991). *People's Parks: The Design and Development of Victorian Parks in Britain*. Cambridge: Cambridge University Press.

Conway, H. (2000). Everyday Landscapes: Public Parks From 1930 to 2000. *Garden History*, 28(1), 117–134.

Dagger, R. (1981). Metropolis, Memory and Citizenship. *American Journal of Political Science*, 25(4), 715–737.

Davies, A. (1992). *Leisure, Gender and Poverty: Working Class Culture in Salford and Manchester 1900–1939*. Buckingham: Open University Press.

Editorial (1902, 25 September). *Manchester Evening Chronicle*, p. 2.

Fraser, D. (1979). *Urban Politics in Victorian England: The Structure of Politics in Victorian Cities*. Basingstoke: Macmillan.

Garrard, J. (1983). *Leadership and Power in Victorian Industrial Towns 1830–1880*. Manchester: Manchester University Press.

Golf in the Parks (1936, 12 May). *Manchester Guardian*, p. 13.

Hannikainen, M. (2016). *The Greening of London, 1920–2000*. London and New York: Routledge.

The Health Resorts (1926, 15 July). *Manchester Guardian*, p. 12.

Henry, I. (2001). *The Politics of Leisure Policy* (2nd Ed.). Basingstoke: Palgrave Macmillan.

Hill, J. (2002). *Sport, Leisure and Culture in Twentieth Century Britain*. Basingstoke: Palgrave Macmillan.

Hollis, P. (1987). *Ladies Elect: Women in English Local Government 1865–1914*. Oxford: Clarendon Press.

Holt, R. (1992). *Sport and the British: A Modern History*. Oxford: Clarendon Press.

Horne, J., Tomlinson, A. and Whannel, G. (1999). *Understanding Sport: An Introduction to the Sociological and Cultural Analysis of Sport*. London: Routledge.

Hull City Council to Allow Sunday Golf (1930, 4 April). *Hull Daily Mail*, p. 15.

Hull Parks Estimates Down £900 (1937, 20 January). *Hull Daily Mail*, p. 10.

Hull's Motorcycle Races (1935, 5 July). *Hull Daily Mail*, p. 15.

Hunt, K. (2012). The Local and the Everyday: Interwar Women's Politics. *The Local Historian*, 42(4), 266–279.

Inglis, S. (2004). *Played in Manchester: The Architectural Heritage of a City at Play*. London: English Heritage.

International Sheepdog Society (2018). History of the ISDS. Available from: www.isds.org.uk/the-isds/history-of-the-isds/the-growth-of-trials/

Jones, S. G. (1990). Working Class Sport in Manchester Between the Wars. In R. Holt (Ed.). *Sport and the Working Class in Modern Britain* (pp. 67–83). Manchester: Manchester University Press.

Kidd, A. (2002). *Manchester*. Edinburgh: Edinburgh University Press.

Lancashire Record Office, Preston, Parks and Recreation Grounds Committee Minutes, CBBU/39.

Liverpool Archives, Liverpool, Parks and Gardens Committee Minute Books, 352/MIN/PAR/1/.

Manchester Amusements (1908, 7 March). *Manchester City News*, p. 7.

Manchester Archives and Local Studies, Manchester, Parks and Cemeteries Committee Minute Books, GB127. Council Minutes/Parks and Cemeteries/1-53.

Meller, H. (1976). *Leisure and the Changing City 1870–1914*. London: Routledge.

Minister Opens New Parkway (1932, 1 February). *Manchester Evening News*, p. 6.

Mitchell, H. (1928). Women in Public Life. *Newton Heath Free Gazette*, p. 2.

Motor Cycle Races (1939, 11 February). *Western Morning News*, p. 10.

Motor Racing in Hull (1932, 20 August). *Hull Daily Mail*, p. 1.

Pearson, G. (1983). *Hooligan: A History of Respectable Fears*. London and Basingstoke: Palgrave Macmillan.

Pettigrew, A. A. (1926–1932). *The Public Parks and Recreation Grounds of Cardiff*. 6 Volumes. Unpublished.

Physical Training and Recreation Act, 1937, 1 Edw 8 & 1 Geo 6, c.48.

Physick, R. (2007). *Played in Liverpool: Charting the Heritage of a City at Play*. London: English Heritage.

Plan to Lease Parks for Entertainments (1936, 24 August). *Manchester Evening Chronicle*, p. 8.

Princess Parkway (1931–32). *Journal of the Manchester Parks and Cemeteries Staff Society*, Number 6, pp. 31–33.

Public Health Act, 1925, 15 & 16 Geo V, c.71.

Pussard, H. (2004). 'A Relic of Bygone Days'? The Temporal Landscapes of Pleasure Grounds. In C. Aitchison and H. Pussard (Eds.). *Leisure, Space and Visual Culture: Practices and Meanings* (pp. 41–58). London: LSA Publications.

Rodrick, A. (2004). *Self-Help and Civic Culture: Citizenship in Victorian Birmingham*. Aldershot: Ashgate.

Salford Archives, Salford. Parks Superintendent Report Books. L/CS/DR4/1-13.

Savage, M. and Miles, A. (1994). *The Remaking of the British Working Class 1840–1940*. London and New York: Routledge.

Stevenson, J. (1984). *British Society 1914–1945*. London: Penguin.

Theatres' Triple Terror (1926, 18 March). *Manchester Evening News*, p. 7.

Theatrical Managers' Association Archives, Minute Books, London.

Tranter, N. (1998). *Sport, Economy and Society in Britain 1750–1914*. Cambridge: Cambridge University Press.

Walton, J. (1983). *The English Seaside Resort: A Social History 1750–1914*. Leicester: Leicester University Press.

Women at Exercise (1936, 25 February). *Manchester Guardian*, p. 13.

Women on Committees (1913, 13 November). *Manchester Courier*, p. 6.

Zweiniger-Bargielowska, I. (2010). *Managing the Body: Beauty, Health and Fitness in Britain, 1880–1939*. Oxford: Oxford University Press.

6 Parks and the Green Agenda

Introduction

Urban parks have an important connection with the environmental movement. Initially established to improve the health of the working classes,
they developed into vital green spaces in the cities of the twentieth century and beyond. Such ideas were informed by debates about the quantity
and quality of open space in the urban environment and concepts such as
biodiversity, conservation and sustainability. The garden city movement
of the early twentieth century sought to combine the urban and the rural
in search of the ideal living environment. This chapter places the urban
park in the context of the early environmentalist agenda and examines
questions about rights of access to green spaces, the contribution of parks
to the greening of the cityscape and their links to other elements of the
green urban agenda, such as street trees and parkways.

Early Forms of Environmentalism: Parks as an Ecology

To what extent were urban parks regarded as distinct environments in
their own right? New initiatives in environmentalism have often been
interwoven with concerns about social justice, stemming from ideas such
as civil rights, suffrage and women's liberation (Nettle, 2014). However,
if we can trace such notions back rather further to the beginnings of the
public park movement in the 1840s, we can observe that ideas about
parks, open space, public health and freedom of access were entwined
from this starting point.

The nineteenth-century city was often a grim and overcrowded place.
However, this often belies the fact that many people were beginning to
become aware of the importance of green space in these unhealthy urban
environments. Much of this discourse emerged in the context of the early
discussions about the establishment of public parks. There was an open
acknowledgement of the significance of the relationship between access
to open space and both physical and moral health. A lot of the evidence
given to the 1833 Select Committee on Public Walks emphasised this

aspect. Dr James Phillips Kay, in his written submission, lamented the fact that health of Manchester's working population 'certainly suffers considerable depression from this deprivation' (1833, p. 66). An 1844 meeting held at the Free Trade Hall in Manchester, which attracted operatives from both that city and the neighbouring Salford, noted that 'mortality is greatest where the atmosphere is worst' (Ruff, 2016, p. 14). It continued: 'what we need to preserve the health of the town is a greater amount of vegetation, open spaces for ventilation, active recreation and exercise, so as to oblige us to breathe the greatest possible amount of oxygen to purify the blood' (Ruff, 2016, p. 14). Although not regarded as such at the time, this is as close to an early statement of the benefits of a green environment for working people as it is possible to find.

Such sentiments were often echoed by local councillors and elected representatives. No doubt there was an element of telling voters what they wished to hear (public parks were notoriously difficult to be opposed to); this also represents a growing sense that at least one of the solutions to the perennial problems of the industrial city was to increase the amount of openly accessible green space. The repeated analogy of parks as lungs or green lungs has already been mentioned as one of the most powerful and recurring images of the need for public parks. The ability to reduce environmental arguments to a single organ of the human body is more than political whimsy; it is a persuasive expression of the impact of clean air on general health. Of course, it should not be overlooked that many public parks suffered from the pollution that pervaded the nearby city and too many were located in close proximity to already-existing industrial zones. The mere existence of a public park was not in itself a solution to the many health problems of the nineteenth-century urban area.

The nineteenth century saw the emergence of a number of pressure groups to increase the amount of open green space in cities and to preserve open space and the natural environment more generally. The Commons Preservation Society was established in 1865 with a view to preserving public footpaths and rights of way. Activists included William Morris and Octavia Hill (later to be involved in the founding of the National Trust). The primary motivation of such groups was to work for the preservation of the natural landscape, often to be found outside of the boundaries of the city. Nevertheless, the emergence of a preservation instinct was to be applied later in urban areas and stimulated debates about public access to open land and the positive effects on the human body from exposure to nature.

The purchase and development of public parks in English cities were fraught with problems and difficulties. To take Liverpool as an example, during the 1860s, much time, energy and effort was expended by the Improvement Committee considering the questions of park development alongside that of commercial concerns and other ancillary issues, such as the provision of roads and railways (LA, Parks and Gardens Committee minutes, Volume 4). Public parks were explicitly acknowledged as part of an overall municipal improvement strategy in a city that was blighted by slums and pollution. However, it is also clear that developing a cohesive plan was far from easy. The provision of public parks was hampered by the lack of a financial and legal apparatus to fund their acquisition and maintenance; parks were not always perceived as an urgent need due to an emphasis on slums, overcrowding and traffic problems. Commercial concerns, such as roads and railways, got a consistently higher priority than public health during this decade. Variable weather conditions were a continual threat to the ability of the parks to generate income, however modest. The parks superintendent for Manchester, W. W. Pettigrew, lamented that 'a dull day or a shower in the morning will make a difference of thousands in the numbers of people visiting the park' (MA, Parks and Cemeteries Committee minutes, Volume 46, p. 23).

The polluted nature of the environment contiguous to the parks of the cities in this study was an ongoing issue. In 1871, a report was submitted to the Manchester Parks and Cemeteries Committee on the effects of pollution on Philips Park, one of the city's three original municipal public parks (MA, Parks and Cemeteries Committee minutes, Volume 1, pp. 514–518). The report emphasised 'the large volume of dense smoke emitted from the various chimneys surrounding the park' and that 'the atmosphere was perfectly clouded by it and the smell of the smoke was stifling' (MA, Parks and Cemeteries Committee minutes, Volume 1, pp. 514–518). Five years later, pollution continued to threaten the environment of the park, with the committee now proposing to meet with the representatives of the two worse offending factories (MA, Parks and Cemeteries Committee minutes, Volume 3, pp. 11–12).

By 1880, the Parks and Cemeteries Committee resorted to threats of legal action against the firm of Richard Johnson and Nephew of Bradford Iron Works (MA, Parks and Cemeteries Committee minutes, Volume, 3, p. 204). The firm, in reply, stated that it was building a new chimney as fast as the weather permitted which would allow them to burn less coal and thus emit less smoke (MA, Parks and Cemeteries Committee minutes, Volume, 3, p. 372). The following year, 1881, saw the publication of a report on tree planting by John Wilson, the head gardener at Queen's Park in the city, another of the three original city parks. White reported that the main barrier to efficient tree planting was continuous pollution and the need to enforce the existing smoke abatement legislation. Among his recommendations to address this problem was a suggestion that careful consideration be given to the species of trees planted in the city. Plane trees should be avoided and ash, Canadian and abele poplar and North American thorns (honey locust trees) should be preferred.

White reported that, of 578 trees planted in 1879 and 1880, some 203 had already died (MA, Parks and Cemeteries Committee minutes, Volume 6, p. 7). Coal smoke was an almost permanent problem in this and other industrial cities. Coal was used by both businesses and house-holds and the impact on the environment of a park could be devastating. Smoke damage resulted in 'oily, blackened vegetation' and often rendered the landscape of a public park as unhealthy as the rest of the city (Mosley, 2001, p. 24). It also affected animal life in the parks—it was estimated that Salford's Peel Park had 6,071 species of birds in 1850 and that only 5 remained by 1882 (Mosley, 2001). It was no wonder that John Ruskin had described Manchester's environment as 'the spiritual home of air pollution' (Mosley, 2001, p. 1).

The Manchester and Salford Noxious Vapours Abatement Association (NVAA) had been formed in 1876 to increase the pressure on local industry to address the problem of urban pollution in the two cities (Mosley, 2001). The main focus of the NVAA was the effects of coal smoke in particular as Manchester had almost 2,000 chimneys producing this smoke by the end of the nineteenth century (Mosley, 2001). The initial focus was on smoke abatement technology to reduce levels of emissions, but the NVAA often faced a difficult task persuading local businessmen of the need for investment in these technologies.

There were many attempts to marry the best of the urban and rural environments, even in the late nineteenth-century city. The 1887 Allotments Act allowed for local sanitary authorities to acquire land for the purposes of letting it to members of the labouring population. The intention was to encourage the working classes to grow their own food and to develop horticultural skills. Once again, it should be noted that this was not a novel idea. In early nineteenth-century Birmingham, vacant land had been used by the working classes as garden plots called 'guinea gardens', rented for a guinea for one eighth of an acre (Gaskell, 1980, p. 485). Flowers and vegetables could be grown, but many of these gardens vanished with the flourishing of the city and the arrival of the railways in particular. Gaskell (1980) has pointed out that such activities represented the transfer of rural mores to an urban environment, but it is also possible that they had a more practical and less romantic purpose—self-sufficiency even in a growing city landscape was an appealing and ready source of personal pride and individual industry.

The 1890 Public Health Amendment Act had established the principle of planting street trees in cities and towns. The reverence for trees in the urban environment was a legacy of the increase in appreciation of the value of woodland and a desire to preserve it, which began in the 1870s (Simmons, 2001). Olmsted himself, writing in 1870, described trees as the 'permanent furniture of the city' (Brantz and Dumpelmann, 2011, p. 3), which emphasised their importance in the urban landscape. This thinking is very much reflected in the commitment to street trees in the late nineteenth century.

The degree to which this manifested itself varied. There is evidence that different kinds of trees were planted according to the social character of the area. Almost all new streets built from the 1870s were lined with trees. However, this usually applied only to middle-class and wealthier districts. Limes, laburnums and acacia trees were planted in lower middle-class areas, while plane trees and horse chestnuts were reserved for streets with large, detached houses (Kelly, 2012). The Cardiff City Council initiated a policy on the provision of street trees in 1871. The authority was responsible for the upkeep and maintenance of street trees planted by local property owners and for the replacement of those that died (A. Pettigrew, Volume 2, 1929, p. 70). The original planting of the streets was the responsibility of the landowner, with the consent of the Corporation.

In 1883, this policy was amended and the Corporation began to plant the streets with trees themselves, with the homeowners paying the Corporation directly. A series of subsequent plantings around the city were carried out at the Corporation's own expense after a dispute with residents of Richmond Road, half of whom received bills from the Corporation for the planting of street trees and half of whom did not (A. Pettigrew, Volume 2, 1929, p. 71). The decision to plant such trees (at the Corporation's expense or not) was controversial in Cardiff. Many councillors objected to the planting of trees in prosperous areas of the city, whose inhabitants could afford to pay for their own tree planting and to the detriment of the poorer areas (A. Pettigrew, Volume 2, 1929, p. 72). The Corporation eventually resolved to set aside the sum of £100 per annum for street tree planting, under the guidance of the borough engineer (A. Pettigrew, Volume 2, 1929, p. 72). It is also worth noting that, in many local authorities, Cardiff among them, it was the Public Works Committee (or the equivalent) that supervised the work of planting street trees, work that eventually came under the auspices of the Parks Committee.

In some cases, such as in Manchester, the local Parks Committee collaborated with specialist organisations, such as the Manchester Field Naturalist Society (MFNS), in respect of tree planting in the city (MA, Parks and Cemeteries Committee minutes, Volume 11, p. 84). The MFNS took responsibility for selecting and planting trees and shrubs in key locations around the city centre, such as Infirmary Esplanade. The Parks Committee then paid the MFNS.

Plants were frequently used for decorative and celebratory purposes in the nineteenth-century city. Concern was expressed in Manchester in 1891 at the lack of a palm house in the city for the cultivation of plants to meet the demands of various civic events and rituals (MA, Parks and Cemeteries Committee minutes, Volume 12, p. 58). The future opening of the Manchester Ship Canal would require a lot of flowers, which it was felt that the city did not have the capacity to supply. There was no ability to supply the larger ornamental plants required to decorate the town hall and growing demands for the availability of such plants for civic occasions. The issue was raised again in 1898 when demands for decorative plants for the city were increasing (MA, Parks and Cemeteries Committee minutes, Volume 18, p. 41). The Parks and Cemeteries Committee suggested that a joint venture be explored with the Town Hall Committee to pay for a proportion of the costs of a palm house, which was estimated to be about £500 (MA, Parks and Cemeteries Committee minutes, Volume 18, p. 72). The Town Hall Committee's response was that such a matter was 'not within their province', leaving the matter once again deferred (MA, Parks and Cemeteries Committee minutes, Volume 18, p. 79).

Street trees were generally regarded in the nineteenth century as a benefit for the whole community (Gaskell, 1980). They served a social as well as an environmental purpose. Much of this thinking stemmed from an awareness of the relationship between good physical and moral health and the interdependence of one on the other. Moral science emerged from a sense of good moral conduct and its association with one's environment (Driver, 1988). It was this kind of thinking that underpinned much of the early environmentalism that emerged in the nineteenth century and was further refined during the early decades of the twentieth century. The danger of moral contamination was increased through location in an unhealthy and morally dubious (however defined) physical environment. These ideas crystallised during the public health debates that characterised much of the nineteenth century as national government sought to tackle the growing social problems of slum management and the eradication of diseases such as cholera and typhus. It was believed that these diseases spread through miasmic vapours. Successive public health acts introduced in Britain in the years 1848, 1875 and 1925, along with other interim amendment acts (1890 and 1907), forced the issue of public health regularly into the public arena.

Another powerful element to this discourse was the idea of the city as a sick environment, rather than an emphasis on the illness of its individual occupants. The industrial British city of the nineteenth century was characterised by dirty air, foetid mills and teeming slums. The city had become a toxic place whose ill health was affecting its citizens. The key to resolving many public health problems was a two-stage process—caring for those who were ill and improving the urban environment itself (O'Reilly, 2014). The sick city was suggestive of an uncivilised place where diseases circulated freely among the poor. Epidemics of diseases such as cholera emphasised the failure of cities to present themselves as healthy communities and exposed the weaknesses associated with crowded and dirty urban environments (O'Reilly, 2014).

In Manchester in 1878, the Open Spaces Subcommittee was authorised to plant trees in suitable places in the city, including in the grounds of the Royal Infirmary (MA, Parks and Cemeteries Committee minutes, Volume 4, p. 294). Such was the importance attached to this task that the subcommittee was renamed the Open Spaces and Tree Planting Subcommittee in 1879 (MA, Parks and Cemeteries Committee minutes, Volume 5, p. 31). Meanwhile in Liverpool in 1881, Councillor Lunt was still 'continually pressing' for trees to be planted in tubs to enhance the town's decoration (LA, Parks and Gardens Committee minutes, Volume 13, p. 37).

Instead of planting trees in the poorer areas, many local authorities had a scheme for providing window boxes to those living in council housing to brighten their dwellings. In Liverpool, flower-filled window boxes were supplied to all city tenants whose rents were five shillings a week or less in the early twentieth century (LA, Unauthored, 1905, p. 49). A similar scheme was in place in Hull in 1901, with a prize being offered for the best-cared-for window box as an additional incentive (HHC, Parks and Burials Committee minutes, Volume 6, p. 79). Other strategies deployed by the municipality included the leasing of around 10,000 allotment gardens to the better-off working class in Nottingham in 1871 (Gaskell, 1980). Clearly the idea here was to encourage self-help and moral improvement by virtue of the passing on of the middle-class moral imperatives that defined the Victorian period.

This top-down model of improvement was dominated by ideas about the common good and social respectability as virtues that could be transmitted from one social group to another. This could also be glimpsed in action through bodies such as the Commons Preservation Society and the impact of individuals such as Octavia Hill. It was no accident that so many English cities and towns named their earliest environmentally oriented committees 'improvement committees'. The work of these groups encompassed areas such as the provision of parks and public green spaces, street trees, allotments, construction and widening of streets and roads, clearing of dangerous and derelict buildings and the purchase of land to expand the area of the city or town. These committees were really the engines that fuelled the gradual emergence of the Victorian city proper and that laid the blueprint for their shape and subsequent development. Some individuals, such as Manchester's councillor Charles Rowley, made an intimate connection between environmental justice and its role in improving the life of the individual city dweller (Platt, 2005). Ever mindful of the need for state intervention in this process, Rowley was a supporter of concerts in public parks and exhibitions in art galleries and founded the Ancoats Brotherhood in 1878 (whose membership was open to both men and women) specifically for the purpose of bringing art and literature to the working classes.

While writers have pointed out the role played by philanthropists, such as Rowley, in early environmental activism, it is important to remember that many people understood the significance of having access to open green space in urban areas. William Hesketh Lever's model village of Port Sunlight, Cheshire, begun in 1888, established a philanthropic basis for the lifestyle of his workers but one which was designed to be appealing for the inhabitants (Hickman, 2013). The British Ecological Society was established in 1913, the first ecological group in the world (Berry, 2009). It was formed to develop and promote awareness of and research on the natural environment. This group grew out of the 1904 Committee for the Survey and Study of British Vegetation, which aimed to study and document the range of British flora. This activity was indicative of an increase in interest in the natural world, which intensified during the early twentieth century.

Local governments began to create municipal nurseries to grow plants and food both for their own parks and for sale to the public. Manchester's Carrington nursery was one of the largest in the country and an important source of profit for the city. In some cities, such as Leeds, this happened relatively late—in 1925 (WYAS, Leeds Parks Committee minutes, Volume 3, p. 183). Similarly, flowers and fruit grown in the public parks themselves were often sent to those in hospital, as happened in Liverpool in 1894 (LA, Parks and Gardens Committee minutes, Volume 19, p. 201). In the same year, 100 plants were sent to the occupants of Artizans' Dwellings in the city (a form of social housing) to brighten up their homes (LA, Parks and Gardens Committee minutes, Volume 19, p. 201). Rare flowers grown in the public parks were used as an attraction for the public. Hull's Pearson Park had a display of orchids that attracted a 'continual stream' of admirers in 1898 (HHC, Parks and Burials Committee minutes, Volume 6, p. 3).

Public parks were used extensively for the growing of food during World War I and became one of the main suppliers of fruit and vegetables during those years. In both Manchester and Cardiff, the parks superintendent took on responsibility for the cultivation of food in addition to his parks duties. Cabbage and leeks were grown in Roath and Victoria Parks in Cardiff (A. Pettigrew, Volume 6, 1932, p. 8). Already-existing flower beds were used for the growing of more ornamental foods, such as beet and carrots (A. Pettigrew, Volume 6, 1932, p. 8). The growing and selling of food in this way were very successful such that the practice was continued in Cardiff until 1920 (A. Pettigrew, Volume 6, 1932, p. 10). Hay was also generated in the parks by not mowing the grass (A. Pettigrew, Volume 6, 1932, p. 11). Billy, the Victoria park seal, was in peril due to reduced food stocks, and there was consideration of a proposal to return him to the sea in 1917. He survived on half rations and on food scraps thrown by the visiting public (A. Pettigrew, Volume 6, 1932, p. 11).

The establishment of municipal nurseries by many local authorities in the early decades of the twentieth century provided the impetus for a gradual greening of the British city that often spread beyond their urban parks. Leeds City Council considered the provision of such a nursery in 1925 in order to grow trees and shrubs for the city's parks and to supply trees for streets and highways (WYAS, Leeds Parks Committee minutes, Volume 3, p. 183). These nurseries also supplied plants and flowers to the town halls for occasions of municipal celebrations and civic festivals. Liverpool's Parks and Gardens Committee erected a special house in the Botanic Gardens, specifically for the purpose of supplying the municipality with flowers and plants. There were complaints, however, about the amount of damage caused to the plants throughout the period of their loan to the town hall (LA, Parks and Gardens Committee minutes, Volume 12, p. 352).

The oversupply of plants and shrubs from these nurseries was often sold to the general public, providing some income and encouraging the principle of domestic horticulture. Salford's parks superintendent reported that in the summer of 1921, 12,000 such plants had been sold to the general public over a period of just two weeks, realising £240 (SA, Parks Superintendent Report Books, Volume 7, p. 246). Manchester's parks superintendent, W. W. Pettigrew, wrote a book entitled *Common Sense Gardening* (1928) that outlined basic horticultural principles for those parks visitors who wished to emulate for themselves some of the planting schemes that they had observed in urban parks. The enthusiasm for gardening shown by amateurs was increasing during the early decades of the twentieth century. A number of new magazines and manuals on gardening were published from the 1870s as well as rising numbers of horticultural societies and shows, many of which took place in public parks (Constantine, 1981). In Manchester, the local Social Questions Union (a secular workers' organisation) asked the Parks and Cemeteries Committee to make the expertise of their in-house gardeners available to the general public who wished to learn about horticulture. The request was granted (MA, Parks and Cemeteries Committee minutes, Volume 14, p. 99). Many people, of course, did not have the leisure time or the domestic space for such activity, ensuring that it was a hobby confined to the better off.

In becoming public parks, many spaces moved from being totally private to a new form of public access for the first time. Contemporary newspaper accounts questioned the wisdom of the acquisition of such parks as the environment they provided was not necessarily suited to the general public. Some feared the costs involved in the change of use, while others worried that the location of the new park was not ideal for its use by the maximum number of citizens. Much of this response was territorial in nature. When Manchester Corporation contemplated the purchase of the 1,200-acre private estate of Trafford Park for use as a public park in 1893, a local councillor opined that as 'Trafford Park was wedged in the centre of Stretford . . . citizens would have to walk a mile through another territory before they got to the park' (*Manchester City News*, 1893).

The possibilities offered by the environment of a public park were not always recreational. Many city councils saw an opportunity for residential or commercial development of the land, either as well as or instead of a park. In 1896, when Manchester Corporation again returned to the question of the purchase of Trafford Park, it was reminded of these prospects by a signed memorial from many of the city's businessmen, including Sir Elkanah Armitage and Sir James Watts (*Manchester City News*, 1896). The potential flexibility of large, open spaces near to industrialised city centres was a significant component of the decision-making processes of many local authorities at this time, and they were very alive to the consequences for both public health and the public purse.

The impact of social class in the provision of public parks in particular areas of towns and cities is difficult to avoid when considering the issue of parks in environmental history. Many parks were located in more affluent areas distant from city centres, while those areas of the city centre were overlooked. Even by 1939, Manchester councillors were still acknowledging that too many of the city's parks were not located near the most densely populated areas of the city centre (MA, Parks and Cemeteries Committee minutes, Volume 53, p. 104). This was, in part, because of the lack of sufficiently large spaces in these areas, which could facilitate the development of parks. So-called pocket parks, which often lacked any green space at all, were often the only recreational amenity to be found in these areas. Manchester opened recreation grounds of 5,675 square yards at Mount Street and 7,028 square yards at Willert Street in 1881 (MA, Parks and Cemeteries Committee minutes, Volume 5, p. 376). The land for Mount Street came from a donation from the city's Improvement Committee, while that for Willert Street was donated by the Watch Committee (MA, Parks and Cemeteries Committee minutes, Volume 7, p. 25).

The evolution of Mount Street recreation ground is an especially interesting example of the challenges posed by the development and maintenance of even a smaller recreation space. In January 1887, the Parks and Cemeteries Committee gave possession of the ground to the Committee for Securing Open Spaces for Recreation for a two-year period. This organisation had been established in 1880 by Herbert Philips, a prominent Manchester businessman and philanthropist, also responsible for the Noxious Vapours Abatement Association (Kidd, 2006). The committee bore the cost of laying out the park and equipping it with gymnastic equipment, bowling and quoits. Fred Scott, the committee secretary, reported visitors of up to 500 per day and commented on the 'provision of a wholesome outlet for the satisfaction of the craving for physical exercise' (MA, Parks and Cemeteries Committee minutes, Volume 10, p. 100).

This differential provision illustrates one of the biggest challenges for the twentieth-century city—how to escape the commercial logic of providing parks where the land was available and affordable and to find imaginative and creative ways to supply green spaces for all urban inhabitants. The 'spatial segregation by class' of the nineteenth-century city continued to pose problems well into the twentieth century (Platt, 2005, p. 10).

From a design perspective, many prominent parks designers approached their task as a combination of pragmatic facilities and aesthetically pleasing planting schemes. As we have seen, by the twentieth century, it was the city parks superintendent and his team who were responsible for the design and laying out of most of the newly acquired parks spaces, often making use of plants and shrubs from the municipal nurseries. While there was a good deal of variation in terms of the size and location of these new parks, there are remarkable similarities in terms of their design features, suggesting that urban parks had indeed become a distinctive ecology by this period. This consensus began to break down somewhat by the mid-twentieth century, when the characteristic curved pathways of the Victorian public park began to be replaced by straighter walkways and parallel lines favoured by modernisers. In many cases, this was a pragmatic decision; the restricted spaces of many new urban parks meant that walking space was at a premium.

Many urban parks sought to develop their own ecological and conservation features, such as tropical greenhouses or aviaries. These were intended as focal points of interest for visitors but also to reflect the need for the exotic or unusual attractions in parks. Salford City Council considered the provision of an aviary in Buile Hill Park in 1907, conscious of the popularity of such features in parks in other cities, such as Nottingham (SA, Parks Superintendent Report Books, Volume 5, p. 91). Liverpool's Otterspool Park, opened in 1932, was designed to resemble a riverside promenade as it ran along the banks of the River Mersey (O'Mahony, 1934). Uniting a park directly with a river was an attempt to bring together two classical ingredients of a public park—grass and water—but on this occasion, using an already-existing stretch of water.

The idea of the public park as a place for a contemplative stroll, regardless of whether by a waterside, is one of the earliest conceptions of these spaces. One of their most important predecessors, in this respect, was the municipal cemetery. These memorial landscapes shared many features with the public park—undulating walkways, planting schemes and statues. It was not uncommon for many municipalities to manage both parks and cemeteries in the same department, for this reason. As spaces for sombre reflection in a quasi-rural location, parks and cemeteries shared a similar ecology, and their mutual management was more than mere convenience. Both required often-considerable amounts of space and continual investment and upkeep, but both, arguably, made a contribution to the improvement of public and civic health (Scholz, 2017).

The open and accessible environment of the public park often had uses beyond the obvious. Hull's Pearson Park was used as a site for meteorological study and the collection of data. In November 1902, Hull's newly appointed parks superintendent, Harry Bursell Witty, was elected as a fellow of the Royal Meteorological Society, a reflection of his personal interest in this subject (Unauthored, 1903). Hull's coastal location and the nature of the flat and wide landscape made it ideal for the collection of meteorological data in this manner. Eventually, this data was considered to be an important commodity and was sold to external firms in the 1920s (HHC, Parks and Burials Committee minutes, Volume 6, p. 54). The data collected was also published fortnightly or monthly in the *Hull Daily Mail* as 'Meteorological Observations at Pearson Park'. This detailed information provided an insight for readers into local trends and patterns in weather, including rainfall totals, number of hours of sunshine and average temperatures.

The unique ecology of the urban park offered opportunities for the accommodation of unusual animals for public display. Many city councils considered the provision of zoological gardens in public parks to stimulate public interest. Parks had always attracted birds and other wildlife, especially with the provision of water, such as lakes. Some park guides included sections on ornithology to assist the public in identifying the birds (A. Pettigrew, Volume 3, 1931, p. 15). An aquarium was constructed in the botanical garden at Roath Park in Cardiff in 1899 to exhibit unusual species of fish (A. Pettigrew, Volume 3, 1931, p. 16). The zoo at the city's Victoria Park grew out of the original aviary, opened in 1901. Many of the animals were acquired as a consequence of Cardiff being a port and the desire of many sailors to dispose of shipboard pets to a good home (A. Pettigrew, Volume 3, 1931, p. 16). The zoological garden was formally opened at the park in 1909. Rabbits were bred at the zoo for food during World War I, but many of the animals there were affected by food rationing also (A. Pettigrew, Volume 6, 1932, pp. 11–12).

The attractions of a zoo in a public park may seem rather obvious, but many city councils were more wary of this initiative than Cardiff's. While the opportunity to obtain a close view of exotic birds and animals was potentially popular, there was the perennial worry about the cost of such an enterprise. Not only were zoos based on the desire to display mammals to the public but also many were closely connected to ideas about conservation and preservation of rare and endangered species. It must be acknowledged that most park zoos did not have collections that included any especially rare animals so this may have been a more general intention than a specific aim.

Zoos were often considered for public parks but not followed through, as occurred in Liverpool in 1911. The Parks and Gardens Committee considered the establishment of a zoological garden at Calderstones Park (LA, Parks and Gardens Committee minutes, Volume 26, p. 305). Some research was undertaken on the provision of zoos in London and Birmingham and at Manchester's private commercial pleasure gardens at Belle Vue. The information on Belle Vue indicated that the venture was profitable but only because of the existence of other facilities at the site. An expert advised the Parks and Gardens Committee that the space at Calderstones was not extensive enough to accommodate a zoo and that the cost would be circa £40,000 (LA, Parks and Gardens Committee minutes, Volume 26, p. 305). Other cities discovered that legal restrictions on the use of parkland prevented them from instituting a zoo—this was the case at Roundhay Park in Leeds in 1906 (WYAS, Leeds Parks Committee minutes, Volume 1, p. 4). The Parks, Allotments and Cemeteries Committee returned to the idea of a zoological garden for the city in 1936, but the resolution was returned to the committee by the city council, possibly due to fears about the expense (WYAS, Leeds parks Committee minutes, Volume 5, p. 90).

Zoos frequently complemented another regular park feature—the botanic garden, which was often designed to preserve native or rare flowers and plants. The Old English Flower Garden established by W. W. Pettigrew at Heaton Park in Manchester emphasised the cultivation of traditional native English flowers. As Denis Cosgrove has pointed out, this effort often had the effect of creating a distinct hierarchy based on 'unexamined anxieties over identity and normalising moral evaluations' (2003, p. 261). While these flower gardens undoubtedly placed an emphasis on particular plants and planting schemes that influenced domestic gardening, they were popular with parks visitors who adopted many of their ideas in their own gardens.

It is striking how often environmental history neglects the role of the public park and its own ecology. Most of these histories emphasise the rural at the expense of the urban and so contribute to the idealisation of the rural landscape over that of the urban. Public parks and their locales go some way beyond rus in urbe and can therefore be regarded as worthy of study in their own right. Their ecology was often the result of a rather haphazard development impacted on by local economics and political influences, but, measured purely in terms of popularity as sites of leisure, they were consistently successful.

The Garden City and the Re-imaging of Urban Space

Some commentators were already planning and designing a more healthy urban landscape. Published in 1876, Benjamin Ward Richardson's *Hygeia, a City of Health* attempted to redesign the city and to emphasise its potentially health-giving properties. Richardson was a medical doctor and thus his remedies were guided by hygienist principles, which tended to dominate the discourse of many approaches to the resolution of urban problems at this time. Not coincidentally, the book was dedicated to Edwin Chadwick, the instigator of the 1848 Public Health Act and sanitary reformer.

Richardson's work stimulated some public debate on the issue of urban improvement. Instead of addressing the specific issue of sanitary reform, Richardson concentrated on a complete redesign of the city in order to emphasise the fact that good design, he argued, could benefit public health and 'design out' the social problems of the Victorian age. There was also an emphasis on the provision of garden space. Each house had an associated garden and all public buildings were accompanied by green space. A clear connection was being made between public health and the availability of green space for all citizens. Richardson's specifications include detailed descriptions of the styles of the various buildings and the materials from which they were to be constructed, even specifying the colours of the brickwork. House roofing was to be 'all but flat', and the buildings included provision for gardens and for growing flowers (Richardson, 1876, p. 24). The garden square attached to each house was to be supplemented by a children's playground to provide recreational space. Thus, Richardson was transposing the existing elements of the public park into the residential sphere.

One of the most obvious parallels with parks as twentieth-century green spaces is the garden city movement. As Hardy (1991) has commented, this developed from the general urban social reform initiatives of the nineteenth century. However, it is equally important to remember that this was part of a broader shift in thinking about and writing about cities and the role of green space in the urban environment more generally. The emergence of new disciplines such as town planning was beginning to impact on the conception of urban space and to reflect the fact that rural areas were also suffering as a result of the mass movements of people to reside in urban areas. The garden city movement sought resolutions to the problems of both depleted rural and congested urban areas and to utilise the best of both to redefine city life.

The best-known name to be associated with garden cities is Ebenezer Howard. Howard's ideas were based on three magnets—town, country and town-country (the garden city). Howard was a keen social reformist and ideas about self-sufficiency were key to his model. He envisaged, for instance, that the land on which a garden city was constructed had to be community-owned. This form of co-operativism underpinned Howard's belief in social change through design. Later manifestations of the garden city moved away from this conception and placed more emphasis on the physical form of the garden city.

The garden city movement was characterised by the involvement of influential businessmen and professionals, such as Lord Alfred Harmsworth and George Cadbury. The original garden city was envisaged as a network of 'decentralised but inter-related' series of garden cities (Ward, 1992, p. 2), but this did not come to fruition for pragmatic reasons. The later garden village and garden suburb were more practical realisations of this model. The Garden City Association was established in 1899 to enact Howard's ideas, and by 1902, the land for the first garden city at Letchworth had been acquired. Shares in the first Garden City Pioneer Company were bought by WH Lever of Port Sunlight, Lord Harmsworth of the *Daily Mail* and George Cadbury of the chocolate business, demonstrating the strength of interest in the idea (Hardy, 1991). The involvement of prominent businessmen, such as those mentioned, further underlines the commercial nature of the garden city in practice. Once built, both Letchworth and Hampstead Garden Suburbs were governed and run by non-elected company officials and trustees (Meacham, 1999).

A major criticism of nineteenth-century social reformers was that they each focused on a single element of urban life, such as sanitation, slum housing or overcrowding (Meacham, 1999). The upshot of this was that the urban environment as a whole and all of its problems and challenges were not regarded in their entirety. The early decades of the twentieth century saw a transition away from this compartmentalised thinking to regarding cities as organisms whose constituent parts affected the whole.

New public parks could also be a stimulant to the creation of new suburbs. Alkrington Garden Village was built in Middleton, Manchester, close to Heaton Park, the first house opening in 1911. Constructed on garden city principles by developers Sir George Pepler and Ernest Allen, the site covered 700 acres and was designed to hold 12 houses per acre (Culpin, 1913). In 1926, Ernest and Shena Simon donated Wythenshawe Park to Manchester City Council and a new suburb of Wythenshawe was created around it, based on a large municipal housing estate. Its location to the south of the city enabled the parks superintendent, W. W. Pettigrew, to suggest that the new park should be developed as a horticultural centre due to the relatively pollution-free atmosphere, which would facilitate the growth of plants there that could not flourish in more polluted parts of the city (MA, Parks and Cemeteries Committee minutes, Volume 50, p. 42). Sports such as football and cricket would be prohibited to allow for the rural character of the space to be preserved. An educational garden would be established there to teach students of botany and pharmaceuticals as well as the general public, emphasising Pettigrew's continuing commitment to parks as didactic spaces.

A demonstration garden would contain named plants so that the public could learn the names and grow them in their own gardens. In this way, Pettigrew believed that it would be possible 'for even the very poorest people to participate in a joy that usually is peculiar to the wealthy' (MA, Parks and Cemeteries Committee minutes, Volume 38, p. 81). Pettigrew referred to Wythenshawe as a 'City Park' as distinct from a local park, which would develop on 'an ordered basis' (MA, Parks and Cemeteries Committee minutes, Volume 38, p. 81). Wythenshawe Park ended up developing on a more piecemeal basis than Pettigrew intended as it coincided with a period of financial tightening, which he attributed to disastrous income across the city from sports and games in the parks and a consequence of two compensations claims arising from accidents in the parks (MA, Parks and Cemeteries Committee minutes, Volume 51, p. 132).

The estate was originally designed on garden city principles, which is significant as most attempts to use this as a basis for design had been in the south of England at Letchworth Garden City (1904) and Welwyn

Garden City (1920). Similarly, Liverpool City Council used Norris Green Park to develop a new suburb of some 7,000 council houses in 1933 (O'Mahony, 1934). These new developments reignited some aspects of the original questions about the nature of urban parks—who were they really for and who benefited most from them (Douglas, 2013)?

Wythenshawe was to prove both a challenge and a blessing to Manchester—it was an extensive addition to the amount of open green space in the city, but its management and the relationship to the council housing estate were to prove problematic. The city's long-serving parks superintendent, W. W. Pettigrew, retired in 1932 and was replaced by the Salford parks superintendent, John Richardson. In February 1935, the new parks super-intendent presented the Parks Committee with a report on the develop-ment of the entire Wythenshawe estate (including the park). Richardson sought to establish Wythenshawe Park as a major horticultural centre for the city due to its comparatively unpolluted atmosphere, well to the south of the city. His report emphasised the space as a 'city park' as opposed to a purely local park, an interesting distinction that tried to redefine the nature of an urban park for the twentieth century.

Richardson suggested that sports such as football and cricket should be prohibited at Wythenshawe in order to preserve the 'rural character' of the space (MA, Parks and Cemeteries Committee minutes, Volume 50, p. 42). This was to be a place where the visitor could roam freely and picnic under the trees without being disturbed by the playing of games. Richardson planned the creation of a maze, a demonstration garden and an educational garden, all with the aim of increasing the interest of local residents in horticulture. The demonstration garden would be based on a model cottage garden, with all plants clearly labelled so that the public could learn what they wished to plant in their own gardens. In this way, the park could function as a kind of map that could be replicated in the home gardens of visitors (MA, Parks and Cemeteries Committee minutes, Volume 50, p. 43). The educational garden had a more specific focus—the student of botany and of pharmaceuticals as well as the general public.

The Park as Therapeutic Landscape

The original public parks were framed by ideas about the value of access to open, green space in improving public health. Indeed, this remains one of the most consistent arguments for their availability. An explicit connection was made between access to urban open space and physi-cal and moral health. In the 1840s, this rarely translated into the ability to improve mental health also as the correlations were not completely understood. However, the impact of quiet contemplative spaces on psy-chological well-being was beginning to be appreciated. Urban cemeteries had been designed, in part, to allow those visiting graves of family mem-bers the opportunity not only to reflect on their own mortality but also to take advantage of the tranquillity of the atmosphere in its own right. Most cemeteries provided seats, benches and often statues and pathways, all elements to be found later in municipal parks.

The potential impact of a peaceful landscape on a human body and mind was also being appreciated in therapeutic environments, such as hospitals and private asylums, in the nineteenth century. The lessons learned from these places were to be important influences on the design and use made of later public parks. The therapeutic value of the public park was especially significant during wartime. Parks served many useful and pragmatic purposes during World War I, both as places where injured soldiers could recover their strength and for food production on an intense basis. Those parks which had a substantial mansion house within them converted these often underused spaces into temporary hospitals, not merely because of the existence of these large buildings but also due to the positive impact of the wider park landscape on the patients.

Much pioneering work was completed in some parks on the science of rehabilitation and on the development of early forms of prosthesis. The physical environment of Heaton Park in Manchester was available to those recuperating soldiers who were recovering in the park's Heaton Hall during World War I (Figure 6.1). Overseen by the Royal Army Medical Corps' Robert Tait McKenzie, a Canadian who developed the modern science of rehabilitation, the soldiers used the park landscape for exercises designed to strengthen their limbs. Their progress was documented by McKenzie in his book, *Reclaiming the Maimed* (1918). McKenzie combined the use of massage techniques and hydrotherapy with more traditional aerobic-style exercises done in the fresh air of the park. These included knee raises, deep breathing exercises and balance exercises on a series of elevated beams (McKenzie, 1918). This approach to the healing power of the landscape was copied during subsequent conflicts in other countries. At Heaton Park, the parks superintendent, W. W. Pettigrew, supplied recovering war patients with a garden plot and offered a prize for the best display. This was an overt acknowledgement not just of the rehabilitative power of the park's landscape but also of the therapeutic value of cultivating nature. This kind of thinking continued well after the war, when the newly appointed parks superintendent for the city of Salford, John Richardson, commented in 1926 that 'nature is medicinal and restores tone to the body and mind' (SA, Salford City Council Proceedings, 27 October 1926, p. 739). The expertise of the Salford Parks department staff was harnessed to work on the gardens being developed at Nab Top Sanatorium, Ladywell Sanatorium and Hope Hospital (SA, Parks Superintendent Report Books, Volume 2, p. 143).

Post-war, a scheme was enacted in Manchester to train disabled ex-servicemen in horticulture at the municipal nursery at Carrington. The syllabus (drawn up by Pettigrew himself) included the growing of fruit

Figure 6.1 Wounded soldiers exercising at Heaton Park during World War I.

Source: © University Archives and Records Center, University of Pennsylvania.

and vegetables, herbaceous flowering plants, care and rearing of livestock and tillage and manure (MA, Parks and Cemeteries Committee minutes, Volume 39, p. 56, 64). While the epithet 'green' had yet to be applied to any of these activities, we can see the emerging significance of the physical and psychological impact of an interest in and care for the surrounding environment.

The horticultural products of urban parks, such as flowers and fruit, were often distributed to local hospitals. In 1894, flowers and tomatoes grown in Sefton Park, Liverpool, were sent to city hospitals, while 100 plants were dispensed to the tenants of Artizans' Dwellings in Victoria Square (LA, Parks and Gardens Committee minutes, Volume 19, p. 201). Such dwellings were owned by the city and were designed as an attempt to eradicate poor-quality working-class housing (*Boston Evening Transcript*, 1890). The growing of horticultural produce could also afford opportunities for training of youngsters.

Salford's Parks Committee considered the establishment of a nursery and market garden at Culcheth Cottage Homes to provide a training course in horticulture for the resident boys there. The complex, built by Salford Corporation in 1903, provided accommodation for poor children under the supervision of an adult. The fruit and vegetable produce that resulted would be sent to local hospitals, while the trees could be planted in the city's parks (SA, Parks Superintendent Report Books, Volume 11, p. 37).

As we have seen in previous chapters, the tendency to associate parks closely with nature and with the natural environment has a long history. However artificial these spaces were in actuality, there was something to be gained from presenting them to the public as natural and from encouraging them to be perceived in this manner. A continuing contrast between the park and the street or the slum preserved this notion that the park functioned as the opposite of the rest of the urban environment. Of course, its many contrivances were just as artificial as the street or slum, but its very verdancy and emphasis on horticulture helped in creating the impression of the natural.

Concessions were often granted to wounded soldiers visiting the parks—in Cardiff, provision was made for free bathing and refreshments at Roath Lake and free admittance to matinee performances (A. Pettigrew, Volume 6, 1932, p. 5). A number of Belgian refugees in the city presented an oak tree to be planted in Cathays Park as a mark of gratitude to the citizens for their kindness (A. Pettigrew, Volume 6, 1932, p. 5). The importance of green space and the therapeutic effects of nature were appreciated during World War I. Helphand (2006) has documented the use of trench gardens by British, French and German soldiers on the Western Front as a means of humanising the conditions in which they found themselves. In this manner, the green landscape is transformed from being a 'passive setting . . . to an active agent in war' (Helphand, 2006, p. 16).

After World War I, ideas about public health and the open air continued to be refined. The open-air education movement acknowledged the importance of fresh air for schoolchildren, and the conveyance of Oakwood Valley Park to the city of Salford in 1934 prompted local Alderman Rothwell to suggest that the new park be used primarily for educational purposes, such as open-air classes and school camping trips (SA, Salford Parks Subcommittee minutes, Volume 10, pp. 66–67). The importance of fresh air and greenery for mental health continued to be emphasised between the wars. In Hull, a letter writer to the *Hull Daily Mail* extolled the virtues of visits to Pearson and Pickering Parks in 1930 when recovering from a stay in Hull City Mental Hospital (*Hull Daily Mail*, 1930).

Conclusion

Many texts on landscape history neglect or ignore public parks altogether. They tend to focus more on rural environments and on rural landscapes. Even the word 'landscape' evokes notions of the bucolic and unspoilt countryside. In part, this may be explained by the British nostalgia for the rural, which has overlooked the move of population from country to city in the nineteenth century. Thus, the failure to include urban landscapes such as parks in the discourse of landscape history has annulled any opportunity to consider such environments as ecologies in their own right. Whyte has remarked that landscapes can be an important focus for loyalties and symbols of national identity (2002).

It is important for their past and future that we regard public parks as part of the wider landscape history. Viewing them in isolation risks failing to appreciate how influential ideas about rural landscapes were in defining and designing aristocratic and then urban parks. Building on these ideas (while often somewhat utopian) defined the original blueprint for the urban park and its constituent parts, many of which are still observable today. While it is, of course, crucial to value the distinctive elements of the public park, we need to develop a sense of their rural precursors as a means to ensure their future protection. Shortly after the period covered by this book, public parks began a period of gradual decline, at least some elements of which remain. In many cases, these parks were intended to replicate on some level the comforts and securities of our rural ancestors. Their restorative properties and their value as sites for leisure and horticultural pursuits continue a long and proud tradition of British heritage.

Parks were frequently considered alongside cemeteries, zoological gardens and botanical gardens and not as green spaces in their own right. Parks were often planned by borough engineers and surveyors and not by those with horticultural knowledge. The local authority was slow to respond to official reports and suggestions for large-scale redesigning of the city—the so-called ribbon of parks proposed to encircle the city. Pressure from local groups and campaigners was applied, but resistance from local landowners was iniquitous to these developments. The Improvement Committee received a memorial from 8,000 working-class inhabitants about the provision of a public park in Parliament Fields in 1858 (LA, Parks and Gardens Committee minutes, Volume 4, p. 96). Plans and suggestions were considered and reconsidered and postponed often because no agreement could be reached or because a municipal election was due. Special subcommittees were established and re-established and reported in their turn.

What is less easy to assess is the success of urban parks as distinct and unique environments in their own right. Nicholson-Lord has argued persuasively that parks ended up 'bland, blank and monotonous' with little or no variety and tending to standardise the traditional view and features of the countryside, without the topographical diversity (1987, p. 29). In part, this may be a reaction to the late twentieth-century decline of public parks and the failure to invest in their upkeep. However, it must also be acknowledged that many larger urban parks did become rather formulaic, with their formal walkways, boating lakes and tennis courts. This standardisation removed the element of imagination from public parks and enhanced their taken-for-grantedness. The public developed an often-rigid set of expectations for the parks, and the municipality concentrated on fulfilling those expectations, often without question. The next, and final, chapter assesses the totality of the evidence provided by this study and re-evaluates the key themes of the opening chapter.

References

Berry, R. J. (2009). British Ecological Society. In J. Baird Callicott and R. Frodeman (Eds.). *Encyclopedia of Environmental Ethics and Philosophy* (Volume 1, pp. 118–119). Detroit: Macmillan Reference.

Brantz, D. and Dumpelmann, S. (2011). Introduction. In D. Brantz and S. Dumpelmann (Eds.). *Greening the City: Urban Landscapes in the Twentieth Century* (pp. 1–16). Charlottesville and London: University of Virginia Press.

Constantine, S. (1981). Amateur Gardening and Popular Recreation in the 19th and 20th Centuries. *Journal of Social History*, 14(3), 387–406.

Cosgrove, D. (2003). Landscape and the European Sense of Sight-Eyeing Nature. In K. Anderson, M. Domosh, S. Pile and N. Thrift (Eds.). *Handbook of Cultural Geography* (pp. 249–268). London, Thousand Oaks and New Delhi: Sage.

Culpin, E. G. (1913). *The Garden City Movement Up to Date*. London: The Garden City and Town Planning Association.

Douglas, I. (2013). *Cities: An Environmental History*. London and New York: I.B. Tauris.

Driver, F. (1988). Moral Geographies: Social Science and the Urban Environment in Mid-Nineteenth Century England. *Transactions of the Institute of British Geographers*, 13(3), 275–287.

Gaskell, S. M. (1980). Gardens for the Working Class: Victorian Practical Pleasure. *Victorian Studies*, 23(4), 479–501.

Hardy, D. (1991). *From Garden Cities to New Towns: Campaigning for Town and Country Planning 1899–1946*. London: E and FN Spon.

Helphand, K. (2006). *Defiant Gardens: Making Gardens in Wartime*. San Antonio: Trinity University Press.

Hickman, C. (2013). The National Health Society and Urban Green Space in Late Nineteenth Century London. *Landscape and Urban Planning*, 118, 112–119.

Hull History Centre, Kingston-Upon-Hull, Minutes of the Parks and Burials Committee, TCM/2/14-44/6.

Hull's Beautiful Parks (1930, 3 September). *Hull Daily Mail*, p. 3.

Kay, J. P. (1833). Letter to R. A. Slaney MP, chairman of the Select Committee on Public Walks, Volume XV, London, p. 66.

Kelly, M. (2012). Urban Trees and the Green Infrastructure Agenda. In M. Johnston and G. Percival (Eds.). *Trees, People and the Built Environment: Proceedings of the Urban Trees Research Conference* (pp. 166–180). Edinburgh: Forestry Commission.

Kidd, A. (2006). *Manchester: A History*. Lancaster: Carnegie Publishing.

Liverpool Archives, Liverpool, Parks and Gardens Committee Minute Books, 352/MIN/PAR/1/.

Manchester Archives and Local Studies, Manchester, Parks and Cemeteries Committee Minute Books, GB127. Council Minutes/Parks and Cemeteries/1-53.

McKenzie, R. T. (1918). *Reclaiming the Maimed: A Handbook of Physical Therapy*. New York: Macmillan.

Meacham, S. (1999). *Regaining Paradise: Englishness and the Early Garden City Movement*. New Haven and London: Yale University Press.

Mosley, S. (2001). *The Chimney of the World: A History of Smoke Pollution in Victorian and Edwardian Manchester*. Cambridge: White Horse Press.

Nettle, C. (2014). *Community Gardening as Social Action*. Farnham: Ashgate.

Nicholson-Lord, D. (1987). *The Greening of the Cities*. London and New York: Routledge and Kegan Paul.

O'Mahony, M. (1934). *Official Handbook: The Parks, Gardens and Recreation Grounds of the City of Liverpool*. Liverpool: Liverpool City Council.

O'Reilly, C. (2014). Death, Dirt and Disease: Newspaper Discourses on Public Health in the Construction of the Modern British City. *Journal of Historical Pragmatics*, 15(2), 207–227.

Pettigrew, A. A. (1926–1932). *The Public Parks and Recreation Grounds of Cardiff*. 6 Volumes. Unpublished.

Platt, H. (2005). *Shock Cities: The Environmental Transformation and Reform of Manchester and Chicago*. London and Chicago: University of Chicago Press.

Real Estate Matters (1890, 29 September). *Boston Evening Transcript*, p. 5.

Richardson, B. W. (1876). *Hygeia: A City of Health*. London: Macmillan.

Ruff, A. (2016). *Manchester's Philips Park: A Park for the People, by the People Since 1845*. Stroud: Amberley Publishing.

Salford Archives, Salford. Parks Superintendent Report Books. L/CS/DR4/1-13.

Salford Archives, Salford. Salford Parks Sub-Committee Minutes, L/CS/COSSA/AM.

Scholz, M. (2017). Over Our Dead Bodies: The Fight Over Cemetery Construction in Nineteenth-Century London. *Journal of Urban History*, 43(3), 445–457.

Simmons, I. G. (2001). *An Environmental History of Great Britain: From 10,000 Years Ago to the Present*. Edinburgh: Edinburgh University Press.

Trafford Park (1896, 6 June). *Manchester City News*, p. 4.

Trafford Park Estate (1893, 4 February). *Manchester City News*, p. 8.

Unauthored. (1903). Proceedings at the Meetings of the Society. *Quarterly Journal of the Royal Meteorological Society*, 29(125), 54–55.

Unauthored. (1905). *Illustrated Handbook to the Parks, Gardens and Recreation Grounds of the City of Liverpool*. Liverpool: George J. Smith and Company.

Ward, S. (1992). The Garden City Introduced. In S. Ward (Ed.). *The Garden City: Past, Present and Future* (pp. 1–27). London: E & FN Spon.

West Yorkshire Archive Service, Leeds. Leeds Parks Committee Minutes, LLC47/1/.

Whyte, I. (2002). *Landscape and History Since 1500*. London: Reaktion Books.

7 Public Parks, Leisure and the Popular Imagination

Introduction

This chapter addresses the position of the public park in the wider context of public and municipal leisure provision during the early twentieth century. This period also marks the beginning of the end of the innovations of the municipal authority, so influential in creating and maintaining the urban landscape since the Victorian era. With the increasing centralisation of powers over health policy after World War II, municipal governments eventually lost their influence as local leisure providers.

Returning to a revised set of key themes and issues of the opening chapter, the conclusion demonstrates the resilience of the urban park as a microcosm of civic life in the late nineteenth and early twentieth century. It emphasises the multifarious and contested nature of public open space and the varied mechanisms by which such spaces were developed, used and altered over time. Some of the early themes, such as public health, have now been subsumed into more general topics (civic and urban culture) due to a lack of consistent evidence for their success.

The real significance of the public park for its users lies in its symbolic importance. As observed at the outset of this study, parks have attained in the public mind the status of unimpeachable spaces in crowded, polluted and congested cities. While they have undergone many changes since their inception, they quickly established themselves as valued and much-loved places for a diverse range of activities. The public clamour for this establishment, followed by their often-controversial usage, has formed and shaped their prospects and how historians have viewed them.

Civic and Urban Culture

One of the most defining characteristics has been the uneven development of public parks, which presents many problems when studying their significance. The sheer number, size and variety of functions mean that attempts to study them in detail are difficult. From the vast flagship parks to smaller pocket parks and spartan recreation grounds, there is no one definition of a public park and no one manner in which they have been developed, funded and supplied with resources. Perhaps the single most important feature of all is the enthusiasm with which the public has embraced them. This demonstrates the power of the public park in the public mind.

There is a tendency for the early years of the Victorian parks to be viewed as a kind of 'golden age' as it has been for many of the social innovations of this period—such as public libraries, art galleries and museums. Parks are frequently aggregated with these institutions when considering the history of middle-class moral imperialism and attempts to 'improve' the intellectual life of the working classes. However, we must exercise some caution here. Parks were also a focus of much of the resistance of the working classes to these top-down attempts to impose a kind of narrowly focused definition of improvement, which invariably discouraged political and/or religious controversies (Rose, 2001). The archival research on which this text is based demonstrates that parks were often the kinds of public spaces in which political and religious issues could be and were debated and where these issues could ignite and focus public opinion formation.

This aspect added a new dimension to the prevailing middle-class civic culture of the late nineteenth century and provided the impetus for the gradual widening of that culture during the early decades of the twentieth century. Some aspects of public parks were clearly intended to inspire their visitors. This is especially true of the gardening and horticultural displays. While some were aimed at the kinds of extravagant creative expression that were beyond the average gardener, such as floral clocks, the mere presence of more common border plants and flowers was intended to indicate to all with an interest in such things that amateur gardening could be carried on with very little effort and resources. The practical skills of parks administrators such as William Wallace Pettigrew emphasised the possibilities available to those who aspired to create nominal gardens and introduced the sale of surplus plants to the general public in order to stress the ease with which one could acquire basic horticultural appreciation.

While these attempts to address civic and urban culture more generally can be viewed as a logical extension of the Victorian impetus of active citizenship and social responsibility, elements of such mutual improvement have also been apparent in working-class culture of this time (Rose, 2001). These efforts were not always an effort to merely emulate middle-class mores but genuine and independent working-class struggles for self-actualisation. The very openness of the public park space and the diversity of activity that could be provided (both officially sanctioned and otherwise) made these spaces uniquely contested and open to interpretation. In short, they could mean whatever their users wanted or needed them to mean.

This emphasises the importance of civic and urban culture in its widest sense—not that which merely pertained to formal civic occasions and rituals but a form of culture in which the public in general could participate and create for themselves. While some of these elements were unwelcome to some parks users, they were not without value. This new type of public space inaugurated new ideas about behaviour in these spaces, and new norms emerged that were not simply responses to predominantly middle-class values. Public parks were not only sites of learning about trees, flowers, planting, public display and the importance of fresh air in the urban environment but also sites of contestation about civic values, authority, sexual mores, deviancy and criminal activity.

It is difficult if not impossible to separate parks from the municipal context in which they were developed. They were highly susceptible to civic priorities and agendas. While many began in the rather narrow context of improving public health, this initial aim was broadened considerably to encompass more pragmatic purposes, such as the provision of sporting facilities. Reading the minutes of Liverpool's Improvement Committee during the 1860s reveals a municipal authority attempting to address many social and economic issues at once. Parks were often quite low on the civic agenda. A special meeting of the Improvement Committee in February 1860 commented that while parks and boulevards would benefit public health, they were not 'a matter of absolute necessity like that of rendering our thoroughfares adequate to the traffic carried on them' (LA, Improvement Committee minutes, Volume 4, p. 126). There was a general acknowledgement that parks would need to be funded by the imposition of an improvement rate with which the public would have to agree. An inquiry to the City's Finance Committee revealed that it would not be possible to develop parks without such a rate being levied (LA, Improvement Committee minutes, Volume 4, p. 129). The commercial priority of providing an adequate transport infrastructure for the city took precedence over the development of parks.

Parks were also very much the product of those parks committees that managed them. Reflecting on her experience as a committee member, Manchester suffragette Hannah Mitchell remarked that she had not found it to be as 'congenial' as she had expected (1968, p. 207). She described her aspirations for parks to be places that prioritised trees, flowers and resting places, whereas her fellow (male) committee members were more committed to sporting provisions. 'Chairs set on concrete set around the bandstand was not my idea of music in the park', she commented (1968, p. 207). However, as we have seen, the development of more sporting activities in public parks was less about gender and more about making pragmatic, income-generating use of the space and balancing these considerations with a continuing commitment to maintaining planting schemes and more restful areas.

Parks had to be responsive not only to local priorities and agendas but also to national government legislation and schemes. In part, their flexibility as spaces served them well in this context. National legislation, from the 1870 Public Health Act to the Unemployed Workmen's Act (1925) and the Physical Training and Recreation Act of 1937, impacted on how parks were perceived and used and provided opportunities for innovation and repositioning.

Normative Behaviour in the Urban Environment

An important element of this was the ability of the social space of the public park to accommodate people from various social and economic classes in the same place and for them to all find meaning in their own experience. Despite formal attempts to regulate and standardise public behaviour in the park, there was no mechanism strong and enforceable enough to challenge all of the normative behaviour on display there. Indeed, parks were sites where some of the most controversial and criminal behaviours were reasonably common. This presented a threat to certain parks visitors, such as women and children, in particular. New public spaces such as parks therefore gave rise to new forms of public behaviour, not all of which were positive.

This research has uncovered extensive evidence that parks were places where paedophiles and other criminals actively preyed on victims and took advantage of the fact that these spaces were so attractive to the general population. There was a high degree of awareness of these threats among those who patrolled, administered and managed the parks, yet these kinds of behaviours were unable to be fully eradicated. The prevailing mores of the time often prevented explicit warnings being issued about individual parks or particular areas within larger parks. Parks could also be used to conceal births, to camouflage the bodies of dead babies or children or to find the privacy so often denied to couples in the domestic environment. The existence of quiet and private spaces and concealed areas even within larger parks enabled opportunities for people to attempt and commit suicide and to gratify many kinds of behaviours that were impossible to regulate or predict.

Public parks as controlled and regulated spaces were, therefore, a failure. Nicholson-Lord's assessment of them as 'thinly disguised instruments of crowd control' does not seem to adequately reflect the reality of the evidence of this study (1987, p. 28). Attempts to regulate and control park visitor behaviour continued to form part of the standard approach of public park administration. Yet, it remains clear that no scheme of regulation could cover all eventualities and the authorities at least tacitly, if not overtly, acknowledged this. So, was this just regulation for regulation's sake? This seems not to have been the case. There is consistent evidence of a public demand and expectation of a regulated environment even if this was a practical impossibility. Newspapers regularly printed letters from park users asking for rules to be implemented and enforced about many kinds of unwanted behaviours—cycling, driving speeds, playing of sports and music and the holding of political meetings—all of them excited public responses and opinions.

When Jane Jacobs described public parks as 'volatile' spaces, this seems to have a generally negative connotation (1964, p. 99). However, it could also be argued that this function of volatility could have its virtues. The sheer flexibility of the space was part of its attraction for the public. It drew people to go there because it was ill defined and problematic in terms of control and regulation. Both old and new forms of public behaviour were demanded there, and people were able to negotiate the spaces for themselves and in accordance with their own behavioural norms, whatever those may be. This might seem overly optimistic, and there is certainly plenty of evidence in this text of formal attempts to police, organise and enforce kinds of behaviours which might be recognised as middle-class. Regardless of whether those attempts were successful (and they certainly were varied), parks represented a valuable and valued asset to their many visitors and users.

While, in many respects, urban parks did little but reproduce the social and economic inequalities apparent in the wider environment, there is also a sense in which they provided a space in which new modes of social and public behaviour could and did emerge. They were often one of the few spaces in which different social classes could inhabit the same place. In practice, this was less about the working classes learning new modes of behaviour from their social betters and more about a range of people from different social and economic backgrounds having the opportunity to explore a leisure space on more or less equal terms. Without overstating it, it gave the urban working classes the chance to experience and to experiment with mutual self-improvement.

As Rose (2001) has pointed out, this mutual improvement took many forms and is conceived of mostly as activities that took place indoors in libraries or religious buildings. This grassroots struggle for improvement could also be conducted outdoors in an urban park and often without any financial outlay. Contemporary photographs of parks visitors provide important clues as to the social and economic mix of people in an urban park, although almost invariably at the weekend. While some commentators of the time remarked on the presence of their social inferiors, there is little evidence of any major disruption or formal detriment to anyone's enjoyment of the space as a result.

Platt has written of 'moral environmentalism', in which immoral environments breed immoral acts (2005, p. 10). However, it is not quite correct to ascribe morality to environments, which, as we have seen, can be designed with one purpose in mind, only to be used for another. In this sense, urban parks really were and are the spaces of the people as their behaviour once in them was often unpredictable and contradictory. It could be said that 'conflict and negotiation may occur as a result of the collision of discourses' (Winchester, Kong and Dunn, 2003, p. 100).

Public parks thus represent a multilayered canvas on which diverse social and political issues could be contested.

Public Leisure and Spatial Segregation

One of the most significant features of the public park was the gradual emergence of spatial segregation. This was an element of the earliest parks, with areas set aside for strolling among flower beds, bowling and bandstands. Even before the existence of the first formal public parks, spatial segregation was a feature of many public spaces—private gardens needed a key to provide access to visitors, and pleasure gardens also regulated access to those of particular social classes or charged a fee, which was designed to deter the poorest.

This spatial specialisation continued to develop during the early twentieth century with spaces developed for cricket, tennis, boating lakes, dancing and theatrical performances. It was the very versatility of these parks that marked their hold on the public imagination. They could be whatever people needed them to be and were able to be adjusted according to public fashion and mood. It is in the larger, flagship parks that this spatial segregation is most evident. The park walls, gates and formal entrances all functioned as effective boundaries—these were enforced by the closing of the park gates, usually at dusk, and thus the exclusion of those who wished to remain inside. The different parts of the park could be clearly delineated—many parks had maps at the entrance, directing visitors to the various spaces.

These developments could be perceived as part of a wider repositioning of parks that took them away from the original 'rational recreation' principles of their founders and sought to emphasise more general commitments to a broad range of public leisure activities. This runs the risk of conceiving of public parks as confined to the larger flagship parks that could accommodate a variety of activities easily in their acreage.

The smaller recreation grounds (or 'pocket parks') that were so much a feature of the cities in this text were an early twentieth-century response to the issue that many of the larger public parks were located far from the most overcrowded areas of the city and offered no respite to those who most needed them. Many of these grounds were simply too small to benefit from any kind of spatial segregation—most had limited or no greenery at all—and offered only some playground equipment to attract and amuse children. While the large parks got the most attention from visitors and in terms of facilities, these smaller recreation grounds should not be overlooked when assessing their impact on public leisure. They were often the only source of recreation for young children in congested cities and provided a valuable outlet for their energies. They were also continually invested in as the equipment was upgraded and maintained. Evidence from Manchester demonstrates that while many of these smaller recreation grounds brought in no money at all (many had donation boxes to allow for the collection of money), they had an annual cost ranging from £70 at Monsall Road Recreation Ground to £450 at Willert Street Recreation Ground in 1912 (O'Reilly, 2009).

In large part, these smaller recreation grounds reflected a period of continual development of the cities in this study. Some of the grounds were located quite near to larger parks, such as Willert Street in Manchester, mentioned earlier, which was close to two of Manchester's earliest public parks, Philips Park and Queen's Park, but was developed anyway due to the commitment to supplying open spaces where land could be made available. These smaller grounds were ideal leisure spaces for those who wished to supervise their children's leisure activities or who did not have the means to access the resources at the larger parks. Their main beneficiaries were women and children, demonstrating a new commitment to the public leisure needs of these often neglected groups. However, there is little doubt that many residents of overcrowded and polluted cities did have not easy access to open green space. To take just one example of an autobiography from this period, William Jones recorded his childhood games in central Manchester in the early decades of the twentieth century and makes no mention of visiting a park but of amusing himself playing in abandoned buildings and in the tunnels of old kilns (Jones, 2006). While it is tempting to argue that leisure was now a right and not a privilege, it was not yet extended to all (Jones, 1994). This suggests that not only had the social control model of leisure broken down but also it had not yet been replaced with anything more nuanced.

However, it is notable that spatial segregation also operated in a wider sense than just inside a park. While the larger, flagship parks were often located in or near to suburbs or to more affluent areas, smaller recreation grounds were found in more congested areas or not at all. The poorer districts of Manchester, such as Hulme and Ancoats, found themselves overlooked despite the best efforts of local philanthropists, such as Herbert Philips (Davies, 1992). Lack of access to cheap public transport effectively placed the larger parks beyond the reach of the poorest. The smaller recreation grounds were more commonly named after their locality or street instead of the wealthy benefactor who donated the land for larger parks, such as Peel Park in Salford. This is not to deny that many large parks were named after their local areas also, but the tendency to reflect the names of local businessmen and urban elites in park naming emphasises Sally Morgan's feeling that 'the power of naming, placing or destroying is crucial to the shaping of the memories from which identities, stories and histories may be constructed' (1998, pp. 103–104).

Differing rates at which public parks developed in the cities under discussion in this volume are also evidence of the varying commitment to these spaces among local authorities at this time. Parks were not always high on the municipal agenda, depending on the needs and demands of the locality, the amount of money to invest (parks in every city were expensive both to establish and to maintain), the political priorities of the local parties and the degree of civic agency in the wider city itself. As Katy Layton-Jones has remarked, 'there is little evidence to suggest that those who provided and managed the nation's public parks in the early years thought of them as commercial assets' (2016, p. 4). Indeed, most parks committees seem to have been acutely conscious of the costs of their undertakings, not just of the acquisition of the space but also of its upkeep and maintenance. Donations of land by wealthy local benefactors were often more of a curse than a blessing as the park then had to be laid out and developed over a long period of time. Those who were generous locally were often less so at a national level. Robert Peel, when prime minister, was approached for a contribution to the development of Manchester's parks and offered only £3,000 (Layton-Jones, 2016).

Perhaps the only pertinent observation is the lack of a clear and distinct pattern of these kinds of developments in most urban areas. There was, therefore, no 'golden age', just a series of periods of adjustment and often haphazard developments, predicated on public demand for various sporting and leisure activities, many of which were subject to transient fashions of the time. There was perhaps more negotiation and compromise involved in their development than conflict and opposition—parks are, after all, difficult to be opposed to (at least publicly).

Urban Parks, Public Health and the Popular Imagination

In 1921, at a meeting of the Lord's Day Observation Society, the Conservative MP Sir Herbert Nield exclaimed that people in Britain had gone 'recreation mad' (Graves and Hodge, 1941, p. 103). While undoubtedly intended to be hyperbole, Nield was reflecting a society that had embraced public leisure. However, this had been a gradual and rather uneven process, as this study has shown. Healthy activity and physical fitness were now the responsibility of the individual rather than being imposed by the middle-class moral imperialists of 'rational recreation'. Many people, such as women and children from poorer urban areas, were no better off in terms of access to recreational facilities than they had been in the 1840s, leading to the emergence of what might be termed a 'leisure divide' between those who had both the time and the access (and the finances) for leisure and those who had not. It was the versatility of the parks as spaces that formed the great part of their appeal for their users and not the ability or desire to impose a form of social control. As Stedman Jones has argued, to do so is to reduce studies of leisure and recreation to a polarised and polarising 'arena of struggle' (1977, p. 170). Instead, it is the nuances inherent in the various uses of leisure spaces such as parks and their multitudinous meanings for their visitors that provide evidence for their hold on the public imagination.

The dichotomy between the open, airy public spaces of the park and the more secluded private spaces remains one of the least explored features of the urban park. The primary research presented in this study reveals that parks were often spaces of fear and criminality, where vulnerable children and women were preyed on and where serious crimes could be both committed and concealed. Some aspects of their management and their role as employers remain frustratingly elusive. The voices and opinions of those who worked long and hard in these parks are all too often missing from archives and official accounts. Diligent parks supervisors and managers kept detailed records of the management of these spaces from which it is often possible to infer at least the main duties and requirements of these jobs and the attributes that were welcome and not so welcome, but the thoughts, feelings and views of these men (as they mainly were during this period) are all but absent.

What is abundantly clear is how popular and successful these spaces were for the general public. Many of these cities were characterised by heavy industry, overcrowding and pollution, and the haven afforded by even the smallest green space was precious to many. It is in this fact that we should invest our appreciation for public parks and their survival over time. They were not established purely as a response to a rapidly urbanising society but as a complex series of attempts to resolve the urban/rural dichotomy and a rejoinder to the traditional pastoral ideal (Malchow, 1985). The sheer variety of reactions to the Victorian city is startling—urban dwellers who had the means tried to escape the city periodically by taking day and weekend trips to the countryside or the coast, while the wealthier tried to bring nature to their homes by building villas in the suburbs, surrounded by gardens that recreated the rural past.

The urban park united two elements often perceived as each other's opposite—the city and nature (Brantz and Dumpelmann, 2011). In so doing, nature was altered by being reproduced in an urban space—thus the importance of the study of the 'social construction' of nature in the city (Brantz and Dumpelmann, 2011, p. 11). Urban parks are a vital component of this. One of the principal objections to designed spaces such as parks was their allegedly 'sterile' and often standardised nature—that is, that they were unnatural. The rural exercised a certain romantic fascination for urban dwellers, and we can note this tendency, which continued well into the twentieth century (and beyond it). Much of this was fuelled by how parks were written about and depicted to their visitors, and it was these depictions that contributed to the significance of public parks in the popular imagination.

The visual iconography of the public park can help us to understand not only what was represented but also how it was represented. The images chosen reflect particular priorities and civic expectations. The images are a form of social documentary and offer an opportunity to examine the emergence of a heritage gaze to complement Urry's concept of the tourist gaze—'the socially organised and systematised gaze of the tourist' (1990, p. 1). Like the tourist gaze, the heritage gaze is not homogenous and reflects the fact that many people can view the same image in different ways. Spaces of leisure become transformed into objects such as postcards and guidebooks that are then transferred between people. Again, like the tourist gaze, the heritage gaze is not neutral but privileges some kinds of images over others, thus establishing a dominant ideology about how parks are represented. The process of the selection of images is not overt and visible but insidious and legitimises some interpretations instead of others (Pritchard and Morgan, 2003).

By publishing postcards, guidebooks, official handbooks and other printed ephemera, each city council was choosing how to represent their parks and to whom. They were taking advantage of an increasingly commercial world of leisure opportunities and were struggling to make their voices heard above those of private entrepreneurs. They were also using the means at their disposal to promote their cities and their amenities and to present themselves as providing value for money. This rhetoric often underpinned the text of parks guidebooks and handbooks. Careful note was made of how much was being invested in the parks and of the variety of facilities available at reasonable prices.

Many of the images of parks produced during the late nineteenth and early twentieth centuries reflected civic priorities and expectations. These did not remain static over time. As new editions of guidebooks and handbooks were published, the text and its illustrations could be revised to accommodate new facilities and new emphases. The visual cues could thus have a dual function—as a communication medium and as a promotional medium—ensuring that any monies invested could prove valuable in terms of a return. Landscape and its features were thus rendered as commodities in themselves and offered for consumption by the general public, whose rates had paid for their development and maintenance. The constant historicising of the heritage landscape was a sign that for the civic authorities these landscapes and their meanings were a source of tradition and a means of organising the collective urban memory of place and parkland.

Romantic descriptions of urban parks were common in the pages of local newspapers during the period under consideration. 'Quiz' writing in the *Hull Packet and East Riding Times* in 1876 describes how 'lazily lolling upon the green sward, I watch with half-closed eyes the fretting ripple of the dark waters as a slight breeze agitates their surface' during a visit to the People's Park (1876). Similarly, an unnamed writer for the *Manchester Courier* wrote of the parks there as 'charming sylvan retreats' (1902). Journalistic licence aside, such writing was clearly intended to stimulate appreciation for and interest in the contribution of the parks to urban life and to fuel the popular imagination about such parks along several, rather predictable avenues.

The local press was an important mechanism for mediating public opinion on parks. Their letters pages were filled with praise and complaints about parks, and they offered a significant forum for the interpretation and evaluation of the success of the various parks strategies. Parks administrators such as William Wallace Pettigrew were acutely conscious of the need to use the press, as well as a programme of regular public lectures, as a tool to communicate new initiatives to the general public (1937). He himself wrote a regular column on horticultural trends in parks for the *Manchester Guardian*. The local press was only one method of parks mediation—others were the official guidebooks produced every year by the municipal authority, and advertising posters made for trams and buses reminded the travelling public of the attractions of leisure amenities, while the satirical press offered a more critical view of parks provision. These periodicals flourished during the nineteenth century and represented an alternative view of municipal activities to that of the daily or weekly press. Liverpool's the *Porcupine* noted the 'paroxysms of stinginess' (1867) of the local authority when it came to parks for the city, while Manchester's *Free Lance* lamented the lack of parks in poorer areas of the city. It described the 'thousands of smoke-dried children who run pantingly about close, unhealthy streets in search of amusement' (1867).

It is important to remember also that parks were valued primarily for their green space and not for the buildings that often accompanied them. The classical Heaton Hall, which had been designed by James Wyatt for the Earl of Wilton in 1772, was described rather dismissively by the *Manchester Guardian* in 1902 as a 'plain, low building'. Local historian Thomas Swindells called it 'not of interest either to the architect or the historian' and remarked that there was 'little worthy of note to be seen' inside it (1906, p. 6). The use of some of the space as refreshment rooms seemed to offer a solution as to its future function, but not every parks visitor was convinced. William Bagshaw wrote to the *Manchester Guardian* in 1903 to protest the way the hall was being treated. 'Common nails are driven into the walls to hold advertisements', he reported, 'and the marble pillars are pasted with tariff lists'. The arrangements, he felt, 'show a tasteless haste' and 'were rude and incongruous'. Ironically, the same edition of the *Manchester Guardian* contained a report by the London correspondent of the eighth annual meeting of the National Trust for Places of Historic Interest or Natural Beauty, which noted the trust's recent acquisitions.

Even as late as 1915, Heaton Hall was described as having 'few architectural pretensions' (Sullivan, 1915, p. 57). The hall eventually became a branch of the Manchester Art Gallery after its contents were sold at a public auction. Similarly, at Salford's Buile Hill Park, the mansion was disused and the Corporation struggled to find a purpose for it (*Manchester Guardian*, 1902). The existence of such buildings in many formerly aristocratic parks posed a considerable problem for their new owners. Many of the houses were in poor condition and needed substantial amounts of money to be spent on their repair. Thus, many were neglected or only partially used while resources remained focused on the green space.

Purchasing a large space for an urban park was a major undertaking by any local authority, and priorities had to be carefully decided upon. In most cases, the primary aim was to lay out and open the green space first and then develop any further amenities as necessary. Balancing income and expenditure remained difficult. The availability of specific accounts for individual parks is uneven and inconsistent. However, any snapshot that can be gleaned presents a fragile picture. Liverpool's parks had an income of £9,400 in 1918 with an expenditure of £33,000; this had risen to expenditure of £66,600 with a decreased income of £8,900 in 1920 (LA, Parks and Gardens Committee minutes, Volume 26, pp. 265–266). Cities were acquiring more parks and recreation grounds, and those already in existence needed to be further developed and maintained. Municipal financial resources remained troubled, but there is little serious evidence that local councillors regarded parks (openly at least) as a burden.

In Manchester in 1926, Alderman Tom Fox, reflecting on 12 years of change in the city's public parks, wrote that 'at the beginning, the people belonged to the parks, now the parks belong to the people' (*Manchester Guardian*, 1926). In so writing, he was attempting to articulate the change of emphasis from parks to their visitors and an assumed change of agency from the spaces to those who used them. This statement reflects a subtle but significant reordering of the social landscape of park users during this period. Parks were competing against a thriving private leisure and entertainment industry for the attention of an increasingly time-rich and discerning public. Aside from the usefulness of this remark as a piece of municipal rhetoric, it does indicate a sense of an altered dynamic observable in many of the cities in this study, albeit experienced in diverse ways and at differing speeds. People came to parks with varying needs and expectations, not all of which could be fulfilled. Parks were now just one element in a complex landscape of leisure provision, both public and private. Ensuring the 'enhanced health and happiness of the community' in Pettigrew's words had become more challenging as a result, but there remained a determination and a commitment to this ideal for some decades to come (W. Pettigrew, 1937, p. 101).

Conclusion

A broad-based study of wider public leisure sources enables the context in which public attention gradually moved away from parks as major sites of leisure to be understood. While parks remained patronised, often in large numbers at weekends and on public holidays, they were now just one component in a broader public leisure landscape that competed for the attention of the increasingly sophisticated and discerning consumer.

Dance halls, theatres and cinemas all provided an ample supply of privately funded leisure opportunities that appealed to younger tastes and contributed to the slow decline in the popularity of urban parks in the later twentieth century.

Undoubtedly, further research needs to be conducted on the history of our public parks. We need to understand more about their twentieth-century developments in particular, especially the period of their gradual decline since World War II. A better appreciation of how they fit into the wider urban landscape is necessary as well as more regionally focused studies and those that concentrate on parks in smaller towns and villages. The study of parks needs to be approached from an awareness of their value as part of the civic realm and not just as green spaces and recreational environments. We need to study whether and how they worked alongside town and city planning initiatives that emerged during the twentieth century, especially the post-war reconstruction planning of the 1940s and 1950s, and to appreciate the consequences of parks being considered as part of a more planned and scientific approach to urban life (Walker and Duffield, 1983). It is notable how often a consideration of urban parks is missing from many general landscape histories that tend to focus more strongly on the rural. We need to challenge the consensus views about the nature of landscapes, and it is therefore vital that we include urban parks in this in order to contest what Muir has called the 'primacy of the rural' (1999, p. 235).

As Jordan has noted, parks in theory 'made the people happier and therefore better citizens' (1994, p. 86). This study has shown, however, that the reality was more complex and nuanced. Parks were highly contested, divisive and often-controversial spaces. They frequently offered more than they could deliver and represented a serious drain on the public purse. However, they were also loved and valued by their visitors, neighbours, employees and administrators, and it is this element that has persisted over time. It is in the influence that they exerted on the imagination of the general public that we find the true value and benefit of these open, green spaces.

References

Brantz, D. and Dumpelmann, S. (2011). Introduction. In D. Brantz and S. Dumpelmann (Eds.). *Greening the City: Urban Landscapes in the Twentieth Century* (pp. 1–16). Charlottesville and London: University of Virginia Press.

Changes in the Last 12 Years (1926, 15 July). *Manchester Guardian*, p. 13.

Davies, A. (1992). *Leisure, Gender and Poverty: Working Class Culture in Salford and Manchester 1900–1939*. Buckingham: Open University Press.

Graves, R. and Hodge, A. (1941). *The Long Weekend: A Social History of Great Britain 1918–1939*. New York and London: W. W. Norton and Company.

Heaton Park: Public Opening Today (1902, 24 September). *Manchester Guardian*, p. 10.

Heaton Park's Hall (1903, 11 July). *Manchester Guardian*, p. 6.

How Manchester Is Amused (1867, 27 July). *Free Lance*, p. 27.

In the Park (1876, 28 July). *Hull Packet and East Riding Times*, p. 6.

In the Parks (1902, 16 May). *Manchester Courier*, p. 10.

Jacobs, J. (1964). *The Death and Life of Great American Cities*. London: Penguin.

Jones, G. S. (1977). Class Expression Versus Social Control? A Critique of Recent Trends in the Social History of 'Leisure'. *History Workshop*, 4, 162–170.

Jones, H. (1994). *Health and Society in Twentieth Century Britain*. London: Longman.

Jones, W. (2006). *Different Times: A View of Life in Inner Manchester During the First Decades of the Twentieth Century*. Bedford: Authors Online.

Jordan, H. (1994). Public Parks, 1885–1914. *Garden History*, 22(1), 85–113.

Layton-Jones, K. (2016). *History of Public Park Funding and Management (1820–2010)*. Swindon: Historic England.

Malchow, H. (1985). Public Gardens and Social Action in Late Victorian London. *Victorian Studies*, 29(1), 97–124.

Mitchell, H. (1968). *The Hard Way Up: The Autobiography of Hannah Mitchell: Suffragette and Rebel*. London: Faber and Faber.

Morgan, S. (1998). Memory and the Merchants: Commemoration and Civic Identity. *International Journal of Heritage Studies*, 4(2), 103–113.

Muir, R. (1999). *Approaches to Landscape*. Basingstoke: Palgrave Macmillan.

Nicholson-Lord, D. (1987). *The Greening of the Cities*. London and New York: Routledge and Kegan Paul.

O'Reilly, C. (2009). *Aristocratic Fortunes and Civic Aspiration: Issues in the Passage of Aristocratic Land to Municipal Ownership in Later Nineteenth and Early Twentieth Century Manchester With Particular Reference to Heaton Park* (unpublished PhD Thesis), Manchester Metropolitan University, Manchester.

Our London Correspondence (1903, 11 July). *Manchester Guardian*, p. 8.

Pettigrew, W. W. (1937). *Municipal Parks: Layout, Management and Administration*. London: Journal of Park Administration.

Platt, H. (2005). *Shock Cities: The Environmental Transformation and Reform of Manchester and Chicago*. London and Chicago: University of Chicago Press.

Pritchard, A. and Morgan, N. (2003). Mythic Geographies of Representation and Identity: Contemporary Postcards of Wales. *Journal of Tourism and Cultural Change*, 1(2), 111–130.

Rose, J. (2001). *The Intellectual Life of the British Working Classes*. London and New Haven: Yale University Press.

Salford and the Buile Hill Estate (1902, 1 November). *Manchester Guardian*, p. 4.

The Sefton, Newsham and Other Public Parks (1867, 27 April). *The Porcupine*, p. 36.

Sullivan, J. J. (1915). *Illustrated Handbook of the Manchester City Parks and Recreation Grounds*. Manchester: Manchester City Council.

Swindells, T. (1906). *Handbook to Heaton Park*. Eccles.

Urry, J. (1990). *The Tourist Gaze: Leisure and Travel in Contemporary Societies*. Los Angeles: Sage.

Walker, S. and Duffield, B. (1983). Urban Parks and Open Spaces: An Overview. *Landscape Research*, 8(2), 2–12.

Winchester, H., Kong, L. and Dunn, K. (2003). *Landscapes: Ways of Imaging the World*. Harlow: Pearson Prentice Hall.

Index